DIVISIONS
OF
LABOUR

DIVISIONS
OF
LABOUR

Skilled Workers and Technological Change in Nineteenth Century England

Edited by Royden Harrison and Jonathan Zeitlin

THE HARVESTER PRESS • SUSSEX

UNIVERSITY OF ILLINOIS PRESS •
URBANA AND CHICAGO

First published in Great Britain in 1985 by
THE HARVESTER PRESS LIMITED
Publisher; John Spiers
16 Ship Street, Brighton, Sussex
and in the USA by the University of Illinois Press
Urbana–Champaign and Chicago

British Library Cataloguing in Publication Data
Divisions of labour: skilled workers and technological change in
    nineteenth century Britain.
    1.Skilled labor—Great Britain—History—19th century
    2. Technological innovations—Great Britain—History—19th
    century
    I. Harrison, Royden   II. Zeitlin, Jonathan
    338′.06   HD8389
    ISBN 0-7108-0407-5

Library of Congress Cataloging in Publication Data
Main entry under title:
Divisions of Labour.
    Includes index.
    1. Skilled labor——Great Britain——History——19th century.
    2. Technological innovations——Great Britain——History—
    19th century. I. Harrison, Royden. II. Zeitlin, Jonathan.
    HD8390.D58 1985   331.7′94   84-23988
    ISBN 0-252-01189-9

Typeset in 11pt. Times by MC Typeset, Chatham, Kent
Printed in Great Britain by
Whitstable Litho Ltd., Whitstable, Kent

Contents

Notes on Contributors

DAVID BLANKENHORN, M.A. (Warwick) 1978, has worked for the past several years as a community organiser in Massachusetts and Virginia.

ROYDEN HARRISON, M.A., D.Phil. (Oxford), F.R.(Hist.)S., Professor Emeritus of Social History, Warwick. Editor of *Bulletin* of the Society for the Study of Labour History, 1960-1982, *The English Defence of the Commune* (1971), *The Warwick Guide to British Labour Periodicals 1790-1970* (1977) and *Independent Collier* (1978). He is the author of numerous books and articles in Japanese and Italian as well as in English.

MARK HIRSCH, M.A. (Warwick) 1976, is completing a Ph.D. in the Department of History, Harvard University.

KEITH McCLELLAND is completing a Ph.D. in the Department of History, University of Birmingham, and has taught at the University of Bradford, the Polytechnic of Central London, and Bulmershe College, Reading. He has contributed to *History Workshop Journal*, the *Bulletin* of the Society for the Study of Labour History, and R. Samuel and G. Stedman Jones (eds.), *Culture, Ideology and Politics* (1982).

IAN McKAY, M.A. (Warwick) 1976, is Assistant Professor of History at Dalhousie University, Halifax, Nova Scotia.

ALASTAIR REID, M.A., Ph.D. (Cambridge), has held a research fellowship at Gonville and Caius College, Cambridge, and is now a Lecturer in History at Girton College, Cambridge. He has contributed to *Social History* and to J. Winter (ed.), *The Working Class in Modern British History* (1983).

RICHARD WHIPP, M.A., Ph.D. (Warwick) 1983, is currently Research Officer at the Work Organisation Research Centre, University of Aston in Birmingham.

JONATHAN ZEITLIN, Ph.D. (Warwick) 1981, is a Research Fellow and Director of the project on 'Shop Floor Bargaining, Job Control and National Economic Performance' at King's College, Cambridge. He has contributed to the *Cambridge Journal of Economics*, *History Workshop Journal*, and H. Gospel and C. Littler (eds.), *Managerial Strategy and Industrial Relations (1983).*

Foreword

This is a book about artisans which has been produced
within a post-artisanal research culture. As with its predecessors — *Albion's Fatal Tree* (1975), *The Independent
Collier* (1978) and *Policing and Punishment in Nineteenth
Century Britain* (1981) — the majority of the contributors
are students, or former students, of the Centre for the Study
of Social History in the University of Warwick. From its
inception, the Centre was intended as an experiment in
post-graduate teaching and research. It set itself against the
tradition of isolation, miscellany and low productivity which
has for too long been characteristic of too much in this
division of university labour. But to describe it as 'post-
artisanal' is neither to point to a factory, nor even to a
laboratory — although historians might benefit from a
critical and constructive reflection on the research methods
of natural scientists. It is rather to canvass the economy of
numbers where numbers are assembled around certain
concentrated and interrelated areas of research, such as
labour history or the social history of crime or of medicine.
It is to insist upon the importance of creating a scholarly
community in which the post-graduate comes first instead of
last. It is to challenge the bureaucratic convention that
public money can be assigned either to 'Research' (staff) or
to 'Teaching' (students) and to make this challenge tell
through the enterprise of staff/student literary cooperatives
such as this. It is an attempt, however limited and imperfect,
to practise a style of work in which the free development of
each might become the condition for the free development
of all.

   If none of our books has the tight structure that might be

expected to follow from working to a master plan, then it is still to be hoped that none of them appears to have been cobbled together without regard to unity of subject-matter and line of argument. Relative coherence largely depends upon the moment of discovery of a correct and fruitful way of looking at what is going on in seminars. The earlier this happens the better the chances of training up new research which will address itself to a freshly perceived problem and be alert to the need to test newly-detected relationships. One always regrets that the editorial eye was not sharper and surer. In this instance it proved impossible to assemble all the contributors under one roof. They came to the Centre at different times; some to stay for three or four years, some for only one. A majority of them came from overseas and half of them are now resident abroad. Under these circumstances, the editors became more *dirigiste* than either tradition or their own inclinations warranted. They blasted and bullied, they cajoled and commissioned in a most unlovely manner. For this they apologise to the contributors.

To boast that one has produced a book is to invite criticism of what has been put in or left out. Thus, it may be complained that we attend too much to marginally important trades and not enough to key sectors such as textiles, iron and steel or building. There is an undeniable force in this objection. It was in deference to it that we enlisted outside support to supply the study of shipyard workers which appears as the penultimate study in this volume. However, one of our aims has been to demonstrate that the history of the worker has to be written in relation to the history of his work. Nowadays almost everybody pays lip-service to this principle, but there are still relatively few examples of its being applied in practice. In short, close attention to the reconstruction of the labour process makes claims on space. It seemed more important to try and articulate fresh lines of inquiry and interpretation than to wait for them to be more fully tested. It might be a more telling reproach to point out that we should have tried to include more comparative studies of the type supplied only at the end of this volume. Certainly it would have been rewarding to have accompanied the study of the cabinet-makers with one on the shoemakers or tailors, or to have compared the potters with the flint glass-makers. Indeed, work is proceeding at the Centre which would have put this within our reach; and where possible, I have sought to integrate its findings in my Introduction. The response of two trades which are justly renowned for their production of great autodidacts and astonishingly formative of national labour leaders could have been of exceptional interest. We might have used Alexander Campbell, a cabinet-maker by trade; chosen representative of the glassworkers, editor of a great Labour newspaper and leader of the Trades Council in a great ship-building centre to play a similar role here to that which we made Rhymer play at the end of the *Independent Collier*. But it is more agreeable to attend to what we have been doing than to lament the opportunities we may have lost.

Royden Harrison

NOTES

Lancaster, W., 'The Tradition of Militancy in the Leicester ILP', MA, Warwick, 1979 And the work in progress by A. Kershen on the Leeds clothing industry 1850–1914, MPhil, Warwick, forthcoming.

Matsumura, T., *The Labour Aristocracy Revisited: The Victorian Flint-Glass Makers, 1850–1880* (Manchester, 1983).

Reid, H., 'The History of the Furniture Workers from 1833–1956', PhD, Warwick, forthcoming.

Thorn, G., 'Statement Aristocrats and Sweated Labour: Rank-and-File Militancy in the London Bootmakers' Union, 1885–1895', MA, Warwick, 1973.

Whitehead, A., 'The Decline of the Tramping System', MA thesis, Warwick, 1978.

Introduction

Royden Harrison

I

The ambiguity in our title, *Divisions of Labour*, exactly expresses the unity in diversity of our concerns. This is a book about sectional interests and labour processes. It is about 'the skilled' and 'skill'. Above all, it is about the active relationship between the two. It is about how 'skill' considered as an effective social ascription enters into play with 'skill' considered as a hard won set of judgements and dexterities. Accordingly, it links the debate about the 'labour aristocracy' to the developing discussion of the workplace seen as a battlefield where the existing division of labour is to be preserved or re-defined.[1]

The labour aristocracy debate was sparked off by Henry Pelling in a polemical essay.[2] This concept was alleged to be an invention of Marx, Engels, Lenin and their followers, designed to explain away the failed predictions of proletarian revolution. Subsequently, or so it was alleged, this 'mythical' social presence was made to inhabit many institutional mansions, some of them as insubstantial as that great 'historical fiction', the New Model Union. The labour aristocracy was held to be so vague as to amount to a scandal in sociology, or else so precise that it could be turned against those who tried to employ it in history.[3]

Yet few would deny that this debate has yielded rich returns beyond its rewards in terms of undoubted ironies. At any rate, the ironies will hardly be called into question. Marx appears to have explicitly referred to the labour aristocracy rarely: once before he entered into a sustained relationship with its leading representatives, and once after he had broken that relationship off. Engels was inclined to confuse it with mere respectability, or *embourgoisement*,

both much less subtle and interesting phenomena.[4] Lenin
wanted to root it in 'Imperialism', but without offering any
clearly testable account of how it was so nourished.[5]
Subsequently, one of the most accomplished British Marxist
historians tried to refurbish and redeploy the concept the
better to explain events rather than to explain them away.
Yet while Eric Hobsbawm supplied a much needed anatomy
of the concept, by placing a special emphasis on the height
and regularity of earnings, he inevitably invited charges of
economic reductionism which further complicated the problem of his position in relation to that of the 'Three Great
Teachers'. Hobsbawm did not dwell on the peculiarities of
this stratum and he implied that its political role was as
unproblematic and reactionary as Theodore Rothstein and
other Marxists had always insisted.[6] Hobsbawm offered a
bold and learned sketch, but he left himself no room for
reflection about the possible importance of places; of
communities. As for occupations, they were assigned by him
into categories in accordance with whether he took them to
contain larger or smaller 'aristocratic' and 'plebeian' contingents. There was no attempt to conceal the rough, impress-
ionistic and approximate character of this exercise.[7]

It is a nice matter to determine whether the subsequent
work which Hobsbawm himself did so much to occasion has
done more to redress these preliminary shortcomings or to
multiply the difficulties. Exploration of the labour aristocracy's cultural milieu have been conducted at the expense of
blurring its frontier with the mass, or a significant portion of
the mass, of the labouring population, a frontier Hobsbawm
insisted upon as one which was sharply drawn. In short, a
problem has emerged about drawing a line between the
labour aristocracy and the 'respectable' and 'intelligent' part
of the working classes; the portion which minded its
manners and went in for hard work and for 'rational
recreation'.[8] Much attention has been paid to the political
behaviour of the labour aristocracy and to how it was
perceived by the ruling class, but this too has probably had
its costs. For instance, on balance it tells in favour of
narrowing down the 'golden age' or 'classical period' from
Hobsbawm's 1840–1890 to the third quarter of the

nineteenth century. Moreover, this work on 'the political side' exhibits the 'aristocracy' as capable of showing a certain courage, even heroism, which it would be unwise to dismiss as a mere masquerade, a simple matter of false pretences. Only within an exceedingly stark and elementary revolutionary perspective could it be denied that the labour aristocracy – never mind small contingents recruited from its ranks – could be progressive, at least in the context created by its own presence.[9]

Then again, if a number of local or community studies have seemed, at first sight, to confirm that the concept is researchable, they have subsequently given rise to new uncertainties about the exact connotation of the term itself. Thus, in the hands of John Foster it becomes synonymous with pace-setting and co-exploitation as found among his Oldham engineers.[10] With Geoffrey Crossisk, studying Kentish London, it owes nothing to Foster's postulation of control and management from above, but becomes equivalent to some self-contained élite.[11] With Robbie Gray's account of late Victorian Edinburgh, it secures a remarkably convincing character, but once again at the expense of that sharply drawn frontier which it had formerly been required to display.[12]

It must surely be counted a curiosity and great deficiency of this debate that while labour aristocracy has been considered with respect to its economic anatomy, its cultural formation, its political role and its presence within local communities, it has been almost totally neglected in its occupational dimension. One way of reading this book is to take it to be directed to effecting a correction at this point. But, as with the earlier elaborations, it is not self-evident that the results are more confirmatory of the reality of the concept than disturbing to the received ideas about it. As with the earlier attempts to extend the area of research, it is Hobsbawm's 'gulf', his contention concerning the sharply drawn frontier between the aristocrats and the plebeians, which is most obviously being called into question. It is not clear that he needs to insist on this nor that he is still inclined to do so.[13] Undoubtedly the gulf could and did emerge at specific times or in particular places. But if the problem is to explain how the labour aristocracy established its hegemony over mid-Victorian labour as a whole, then insistence on the gulf's reality would constitute more of an obstacle than an aid. If the labour aristocracy had been increasing the size of the differential while real wages had been stationary or in decline rather than tending to rise; if it had not been felt that all labour is respectable for which there is a fair demand; if 'registered and residential manhood suffrage' had not been seen as a mere technicality, then communication — and with communication, leadership — would have become impossibly difficult.

Certainly, our experience suggests that the requirement of such a gulf for Hobsbawm's account must be severely qualified if not discarded altogether. No such gulf emerges in any of the trades or callings which we have been considering. David Blankenhorn's cabinet-makers would have been bewildered had he been able to ask them about their relations with their labourers. Some of them might have offered some account of their relationships with carpenters and wood machinists, with chair-makers and upholsterers, with carvers and French polishers. A few, but only a relatively privileged few, would have been able to discuss how they stood to their apprentices. If there was a polarity here — as in similar consumer trades such as boot and shoemaking or tailoring — it lay between the men producing to order with expensive imported materials and the unfortunate wretches working at the other end of the trade, enslaved to the sweater or the master of a 'slaughter house'. Between these poles one finds an intermediate group trying to make its way despite the pretensions of the 'exclusives' on the one side and the victims of unregulated competition on the other.[14] But the presence of the exclusives is clearly established, not merely as the controllers of the 'honourable' part of the trade, but as men with a distinct sense of pride in their own identity. They referred to themselves as 'workmen of our class'.

The flint or crystal glassmakers are every bit as troublesome to those who want to maintain Hobsbawm's gulf. To be sure, the gaffer or workman, a fascinating synonym, sitting at the head of his 'chair', appeared as the very model of a labour aristocrat. But notice that his relationship with his 'labourers' (the takers-in) was broken by the presence, in between of his 'servitor' and his 'footmaker'.[15] As for the bakers, the reader of Ian McKay's study will discover that distinctions within the occupation are virtually devoid of all importance compared to geographical ones. The standing and respect accorded to journeymen bakers in Glasgow or Edinburgh is to be contrasted, not with the position of their 'labourers', but with the pathetic situation of most of their counterparts working in London or the provinces. Turn to the five towns of the Potteries and bafflement re-enters in the shape of an astonishing maze of interrelated types of product, scale of enterprise and forms of division of labour which conspire to complicate away the simple and the stark distinctions in status. And it is much the same tale once one enters the shipyards. If the aristocrat/plebeian relationship may turn up rather convincingly here, it fails to materialise there.

Nor is this the only aspect of the old 'Theory of the Labour Aristocracy' which seems to run into difficulties. The boilermakers under Knight might appear as labour aristocrats *par excellence* forever exhorting their labourers to remember their place. But such was the irregularity which distinguished their own pattern of earnings, and so subject were they to the insecurities occasioned by pronounced movements in the trade cycle, that their pretentions to aristocratic status must be regarded as suspect if they are judged in terms of Hobsbawm's own criteria. Besides, we must remember that the boilermakers in question here were members of a union rather than of an occupation. Knight was referring to the plater in relation to his helper. As McClelland and Reid show, relations between the blacksmith and his striker or the riveter and his holders-up bore a very different character. These differences can only be understood by looking more closely at the character of the productive process and the size and nature of the gangs employed within it. Moreover, while Hobsbawm was careful not to claim any simple equivalence between being skilled and belonging to the privileged stratum, he plainly expected a high degree of association between the two and between both of them and being organised. But some of our findings, particularly some of those made by Jonathan Zeitlin, suggest a significant number of occasions in which the less skilled or those with merely plant-specific skills might be more secure and better paid than those employed in the most demanding and difficult departments of the work. It was far from being the case that it was always the most skilled compositor who was the most highly rewarded.

Under the weight of these difficulties, some of our contributors are inclined to abandon the concept of the labour aristocracy altogether. They are tempted to conclude that, at best it is redundant in relation to their interests, at worst a complete red herring. For them it either fails to show up or else they find it so unsettled and so incoherent that they would not be able to recognise it if they came upon it. Broadly speaking, the work of these contributors tends to be clustered together in the second half of the book. It may be pointed out that they tend to be concerned with less traditional trades; with the capital goods sector rather than with the consumer goods sector; and with a later period of time. On the other side, while acknowledging the need for revisionism and even rejoicing in it, I would protest against any easy abandonment of a concept which has, directly and indirectly, enriched our historiography. I distrust any appeal, explicit or implicit, to some unnamed and unidenti-fied type of characterisation of classses or of social strata which is supposed to be perfectly coherent and entirely trouble-free since I know of none.[16]

Perhaps I would be more inclined to look beyond particular trades and across a larger span of time and to ask myself how the break in British labour history which is widely, if not universally, agreed to have occurred around 1850, can be described or explained without any reference to this concept.[17] Certainly Takao Matsumura's recent volume in which the power of the gaffer workman at the point of production is seen as being reaffirmed and re-enforced in the wider life of the community raises considerations which need to be further explored.[18] 'Only connect' remains a wise but demanding injunction. Thus, the Hobsbawmians among us also have their criticisms to make. Is it possible to consider the history of industrial relations as comprising nothing more than industrial relations? However, they can hardly deny that if labour aristocracy works, it must be seen to work at the point of production. Hitherto this problem has rarely been tackled in a sufficiently detailed manner or in ways which hold out any prospect of fruitful comparisons. All our contributors are interested in the nature of craft regulations. All of them are concerned with the sources from which they arose and with the circumstances which determined whether they were maintained in the face of attempts by employers to introduce de-skilling innovations or otherwise redefine the existing divisions of labour.

II

Some workmen have usually been found trying to control the supply of, and demand for, their labour power. Most readers will be familiar with what this means so far as the supply side is concerned. It refers to attempts to establish a standard rate of wages which, so long as it is enforceable, will be too high for it to pay the employer to set on the less accomplished or ill-trained men. In practice, this normally requires the imposition of a closed shop which prevents the employer from enlarging the labour force by recruiting non-society men. These restrictions on entry are typically reinforced by insistence upon apprenticeship and by rules governing the number of apprentices who may be taken on. Most obviously the supply of labour can be reduced by limiting the working day and the working week and by reducing overtime and by restricting workloads. There may be attempts, often intermittent and rather half-hearted, to encourage emigration. Less familiar, but by no means unnoticed, are the attempts to maintain or stimulate demand. The employer who attempts to diminish his dependence upon the skilled by substituting new machinery or by replacing skilled men with semi-skilled or unskilled ones, will encounter outright opposition or be compelled to bargain away some part of his anticipated profit through the acceptance of new wage rates and manning levels where the manning level may relate to the qualifications of those employed as well as to their number. More positively, craftsmen may try, in collusion with the employers, to shut out new cost-effective competition whether foreign or domestic. They may imagine that demand can be increased by lobbying local or central government or by encouraging cooperative production. (It would be a mistake to imagine that the charms of cooperative production were confined to expectations of a new experience at work unrelated to expectations of increased opportunities for securing work.)

The interesting problem is not to make an exhaustive enumeration of the devices by which trade unionists have sought to maintain demand and restrict the supply of their labour power, but to identify the conditions which made for their success or failure. Plainly, the nature of their 'skill' is at the essence of the answer. It is equally plain that if skill be no more than 'the ability to do something well', then that answer must elude us.[19] The fact that such a simple definition can be offered is supportive of Braverman's contentions regarding the degradation of work in the twentieth century.[20] But if it is doubtfully correct as a report on our current linguistic conventions, it is certainly false in relation to those of the Victorians. They employed the word within a much smaller range of usage. They would have been astonished by the proposal to extinguish the difference between mere strength and industry and diligence on the one hand, and the capacity to perform tasks requiring a hard acquired combination of judgement and dexterity on the other. For them the skilled were those capable of performing such difficult techniques. Moreover, for them the term was confined to those who were capable of doing those difficult things where they were *worth* doing, i.e. where a specialised type of labour was scarce in relation to the effective demand for it. In the case of the sailmakers, Mark Hirsch supplies us with a vivid reminder of how 'genuine skill' may become an irrelevance, may be denied all recognition, may be reduced to marginality by the impact of technological innovation upon the operation of market forces. He shows how some of his people were able to transfer their skills from the world of boats to that of great entertainments (marquees) or military encampments (tents). But most of them — not all of them — failed in the essential task of getting themselves securely incorporated within a new division of labour redefined by the advent of coal and then oil, iron and then steel.

But where the technical innovation was less dramatic, the 'skilled' were often better able to get themselves incorporated within the new division of labour in privileged ways. Thus, they came to resemble the Cheshire cat whose smile remained after the rest of him had disappeared. Further, as McKay's bakers and Zeitlin's compositors show us, in very different ways, technical innovation does not automatically or invariably diminish the bargaining power of the 'skilled'. On the contrary, it may occasionally enhance it. Thus, as time passes, the relationship between skill and the skilled tends to become ever more problematic and the division of labour ever less 'obvious' and 'natural'. The relationship between what passes for 'skills' and those who are 'reckoned to be skilled' is to be disclosed neither to a naive assumption that there must be a simple one-to-one relationship between them, nor to the cynical view that being skilled has nothing to do with judgement or dexterity but is simply a matter of organisational prowess.[21] As already observed, it is the play between the two, historically considered, which lies at the heart of the matter. Given that the division of labour is to be seen as a battlefield within which managerial prerogative is asserted, resisted and curtailed, what are the main considerations which enter into the outcome? The question is still worth asking even if no answer can be supplied in any particular case without hard empirical work and even if no general answer, no weighting of large considerations is possible, until the exploration of particular cases has been vastly multiplied.

If we start from the side of capital and the ways in which employers may discover that they are vulnerable to the demands of the 'skilled', we will notice the limits set by any given level of scientific and technological 'know-how' to the substitution of capital for labour. We shall attend to the degree of competition, at the level of the firm and the industry, knowing that the relative inelasticity of demand

and the perishable nature of the product will conspire to raise the employer's ability and his willingness to pay. We shall also look at the composition of capital to determine the ratio between labour costs and total costs and, more particularly, at the ratio of strategically well-placed labour costs to all costs. Perhaps there is something in the glass owners' view that depression, rather than prosperity, furnishes the best time to exhibit employer solidarity against the presumptuous claims of craftsmen. But the matter can hardly be settled without reference to the size and distribution of their reserves.

So far as British industry in the nineteenth century is concerned, it is of importance to remember that unevenness and complexity was the rule. The fragmented structure of family-owned firms and the exposure, more and more as time went on, of much of British industry to violent cyclical fluctuations, impeded the development of managerial structures and extensive capital investment. It encouraged employers to favour labour-intensive techniques and to devolve not only production tasks, but some supervision onto relatively cheap skilled labour.[22] In almost every case, there were great differences in the size of firms, more or less sharp sectoral differences and more or less elaborate segmentation of product markets.

These kinds of jaggedness shatter the portraits of 'the cabinet-maker' or 'the compositor', never mind 'the potter', who could be found in any one of four major divisions of the trade. But they also go some way to explain the strengths and weaknesses of craft regulation. Thus, with the cabinet-makers, as with the tailors and boot and shoemakers, organisational power varied pretty directly with which end of the product market was being supplied. Yet even the first and best organised among them were disadvantaged by the absence of any break as significant to the consumer as that between 'pressed-glass' and 'blown glass'. One of the sources of the remarkable power of 'the flint glassmaker of Stourbridge' – and this character too will dissolve before our eyes under closer inspection – lay here. He could successfully oppose the introduction of new furnaces or of 'gadgets' from Newcastle, partly because he could rely upon the discernment of the consumers of the final product. In particular, flint glassmakers were protected from the competition of pressed glass by the requirements of aristocratic and *haute bourgeois* tables where differences, which were trivial from a utilitarian point of view, became of the highest importance exactly for that reason! Those who baked the bread for those tables were evidently not so well placed. They had to face competition from the traditional household economy whether courtly or not. In general, ease of entry into baking was of a far higher order than in glassmaking, even when due allowance has been made for competition coming from the 'cribs', or small pirate enterprises on the fringe of production. As McKay shows, it was only in the great cities of Scotland, where a peculiar conjunction took place between traditional tastes and population densities so as to make technical innovations profitable and to diminish competition, that the masters and their journeymen were able to arrive at practical accommodations among themselves and with each other. Larger ovens and a concentration of labour in bigger bakeries powerfully contributed to trade union organisation and to the abolition of nightwork.

Jonathan Zeitlin's comparison of engineers and compositors not only furnishes us with another example of how craftsmen might succeed in capturing new machines — there were even pockets of organisation among the sailmakers who were able to do that — but of how important were the factors now under consideration. Plainly, it was greater exposure to foreign competition which explained the ferocity and determination of the engineering employers. On the other side, it was competition among the daily newspaper proprietors along with the peculiarly perishable nature of the product, which made them so vulnerable to the power of the 'chapel'.

Approaching matters from the side, not of the relative strengths and weaknesses of capital, but rather with reference to those of labour, first consideration must be given to the nature of the craftsman's 'mystery' and its importance to the total process of production. Plainly, the ease or difficulty of redefining the existing division of labour or of revolutionising the whole process of production, varied enormous-

ly between one trade or industry and another. But beyond
this, the ability to ride technical innovation might be
crucially affected by the imagination, courage, understanding and organisational ability displayed by the craftsmen
themselves. In particular, how they related to neighbouring
or adjacent workmen within the division of labour emerges
as an under-examined problem, although it was crucially
important. Broadly, the choice lay between exclusion and
incorporation; between erecting defences against real or
potential invaders and taking steps to recruit some of such a
besieging force and make it part of the defending garrison.
A calculated mix of these two strategies is frequently
discernable. (McKay's bakers supply us with a third device
whereby the apprenticed journeymen bakers of Glasgow
and Edinburgh are seen protecting themselves against a
surplus of skilled men in their local labour markets by
exporting them to the south.) Finally, the ability of
craftsmen to look beyond the limited horizons of their own
trade and provide practical displays of solidarity with other
trades, could be of high significance.

The point about exclusive versus incorporationist policies
comes out very clearly from Blankenhorn's account of the
divisions among his organised cabinet-makers. (Incidentally,
he also manages to put paid to the conventional wisdom
according to which the cabinet-makers had no history in the
nineteenth century beyond the yearly round of small
business in local societies which were all quite unimpressed
by the New Model.) Sometimes perceptive trade union
leaders might vainly urge the prudence of a more inclusive
strategy on their recalcitrant members, as in the London
boot and shoe trade; occasionally, the less skilled might
themselves decline to be swallowed by their better organised
brethren in the bespoke sector, as in the Leeds clothing
industry at the turn of the century.[23] The flint glass-makers
came about as close to getting it all right as the circumstances of the case allowed. In this trade the gaffer/workman
at the head of 'chair' presided over a diffusion of craft pride
which proceeded through deference as much as through
aspiration, and bound his 'servitor' and his 'foot-maker' to
him. This binding was not merely conjectural since it took

shape in the organisation of the Flint Glass-makers Friendly Society which united about 75 per cent of those working in 'chairs' to the main exclusion of the unfortunate boys who made up the majority of the 'takers-in'. Thus, the danger which might have presented itself in the shape of the ambitious servitor or the audacious foot-makers was well contained.

This being the case, the gaffer/workman could distance himself from a mere glass-bottlemaker who was to be counted a simple workman compared to the 'artist' that he took himself to be. He would exchange solidarities with engineers and skilled building workers, but he was not inclined to pass the time of day with a bottlemaker any more than with a nailmaker. Such people might dwell in Stourbridge, but not in the parts of it inhabited by 'artisans of our class'. The outsiders, as distinct from such people as the glass cutters with whom some affinity was acknowledged and some association allowed, were subject to a rather rigid system of socio-spatial segregation. Inter-marriage between their offspring was highly exceptional.

Among the potters there were workmen who exhibited a comparable sense of craft pride. Richard Whipp introduces us to the man who boasted that to produce his products of 'pure handicraft' it was necessary to have 'a touch as delicate as that of a lady, yet the strength of a navvy'. And certainly in the slip house, in the potting shops and among the firemen, there were to be found contingents of men whose earnings, whose possession of fine clothes, and whose familiarity with the requirements of decorous and sober behaviour, made them as likely for membership of 'the labour aristocracy' as any a gaffer/workman in the flint glass trade. But none of these people stood at the head of a clearly defined division of labour and conferred an appropriate ranking order on others in terms of their proximity to themselves. They were lost in a labyrinth of divisions of production, of tiles and earthenware and china and sanitary products; they were buried under 104 departments within a large company like Wedgwoods, which was cheek by jowl with some miniscule competitor: they were befuddled by subcontracting and beset by women and children whom they exploited when they could and whom they complained about when they could not. They lived, not in a town which could heighten their sense of identity, but within an untidy conurbation where it was virtually impossible to segregate oneself out of sight of dirt, disease and pain. And if the batter and baller on 5 shillings per week had no more in common with the presser sub-employer on 30 shillings than the gaffer/workman had with his taker-in, there was an openness to technical innovation which made the throwers' and the turners' skills obsolete. There was no equivalent of the jigger and jolly to menace the importance and ascendancy of the gaffer/workman.

The complexity of the scene in the shipyards exhibits many of the principles which are being identified. The shipwrights, for example, show how a formerly key group may manage to sustain something of its old prominent and privileged position within an altered state of things. The boilermakers supply a fascinating instance of the use of inclusion and exclusion. As McClelland and Reid show, they pursued the former policy in relation to the 'holders-on' and the 'rivet boys', while adopting the alternative stand in relation to the 'platers' helpers'.

These considerations relating to adjacent or neighbouring work-people are very clearly relevant to Jonathan Zeitlin's comparison between the engineers – who failed to cover their flank – and the compositors who managed to do so.

To sum up, among the leading components of the craftsman's power one expects to find employers made compliant by the inelasticity of demand for their product: made vulnerable by its perishability: weakened by control over a capital of such a composition that substitution for skilled labour is difficult while its costs are not too significant in relation to total costs. One expects to find workmen clever and determined enough to capture new machines and capable of maintaining their privileged position not always through the well enforced exclusions of outsiders, but also through the formation of alliances 'internally' and 'externally'. But these weightless abstractions are worth little compared to hard and careful examinations of craft regulation and its effectiveness in each particular case.

III

The craftsmen whose histories figure in the following pages have had a bad press. From the Right, they have been assailed for their obscurantism and their myopia. From that side they have been seen as the prime carriers of 'the English disease', which consists of a refusal to acknowledge the benign influence of unregulated competition and to submit unquestioningly to its requirements. From the Left, they have been reproached with the selfish pursuit of their own narrow advantages at the expense of that well-articulated class struggle which was to issue in a Cooperative Commonwealth. We have not taken it to be our business to refute such charges any more than to endorse them. Our concern has been to present our people in the round rather than to rescue them from the self-righteous condemnations of posterity.

If we were asked to affix labels of approval or non-approval, then we might maintain that these craftsmen had the faults of their virtues. The dignity and pride which they exhibited in confrontations with their employers was matched by the profane indifference which they often displayed towards the fate of their 'inferiors', particularly women and children. If they sometimes exhibited a heroic solidarity with their fellow workmen, it was almost always done within the discreetly regulated limits of 'the trades': within the confines set by those identified as 'artisans of our class'. If they adopted a more than merely instrumental view of work, they never thought out the relationship between science and skill. If, as will be shown, they sometimes knew how to defy Ure's arrogant assurance that: 'when capital enlists science in her service the refractory hand of labour will always be taught docility', it can hardly be said that they discovered how to turn that maxim round.[24] Indeed, hitherto the workers by hand and by brain have exhibited neither the strength nor the wit required to avert the fragmentation of processes and the sub-division of interests. They have made hardly more progress in this regard than they have in the related matter of transforming the wages question from a sectional into a class issue; from an industrial into a political one.

It has not been our business here to disclose the ways in which our craftsmen transmitted their practices to the conveyor-belt.[25] It has not been our concern to take the measure of these particular 'peculiarities' of the English and weigh them in relation to their contribution to 'the origins of the present crisis'.[26] Of course they belong, with systematic obsolescences and the aristocratic disdain for business and the *penchants* of the practical man, to the penalties of the early start.

Our book then is relevant. But it carries a lesson for the labour movement's critics as well as its adherents. The tradition which we have identified is tenacious. It has shown itself capable of withstanding the imperatives of the professional engineers as well as the scolding of the schoolmasters. The task is to try and help it to transcend its own limitations rather than to get the entire tradition swept away. The case for the tradition is daily being made out for it by its worst adversaries. Recently, a technologist discussing the scope for general purpose robots in industry remarked: 'It is less obvious that robots will be needed to take the place of human beings in most every day jobs . . . To bring in a universal robot would mean using a machine with many abilities to do a single job that may require only one ability.'[27] Thus, capital must not be wasted, but working people may be consigned to monkeydom without a thought.

NOTES

1. More, C., *Skill and the English Working Class* (1980) partially anticipates this conjunction between the debate and the discussion. See in particular his final chapter. His argument was anticipated at certain points by Stedman Jones, G., 'Class Struggle and the Industrial Revolution', *New Left Review*, 90 (1975).
2. Pelling, H., *Popular Politics and Society in Late Victorian Britain* (1968), pp.37–61.
3. The best guide to the literature is Gray, R., *The Aristocracy of Labour in Nineteenth-century Britain, c.1850–1914* (1981).
4. Engels, F. to Marx, K., 7 October 1858.
5. Hobsbawm, E.J., 'Lenin and the Aristocracy of Labour', *Marxism Today*, July 1970.
6. Rothstein, T., *From Chartism to Labourism* (1929).

7. Hobsbawm, E.J., 'The Labour Aristocracy in Nineteenth-Century Britain', in Saville, J. (ed.), *Democracy and the Labour Movement* (1954). This is reprinted by Hobsbawm along with his related essay on 'Custom, Wages and Work-load', in his *Labouring Men* (1964).

8. Tholfsen, T., *Working-Class Radicalism in Mid-Victorian England* (1976). For a plausible questioning of the firm line division between 'roughs' and 'respectables', see Bailey, P., 'Will the Real Bill Banks Please Stand Up?', *Journal of Social History*, 12 (1979).

9. Harrison, R., *Before the Socialists* (1965); Katz, H., *Anglia in Progu Demokracji* (Warsaw, 1965); Smith, F.B., *The Making of the Second Reform Bill* (1966). Cowling, M., *1867* (1967), is an attempt to maintain the territorial claims of *haute politique* against the invasion of labour history. It ends up trying to prove a negative and with a very honest acknowledgement of being otherwise completely baffled.

10. Foster, J., *Class Struggle and the Industrial Revolution* (1974). But in his contribution to Skelley, J. (ed.), *The General Strike* (1976) he appears to have switched his ground.

11. Crossick, G.J., *An Artisan Élite in Victorian Society* (1978).

12. Gray, R.Q., *The Labour Aristocracy in Victorian Edinburgh* (1976).

13. Conference Report, *Bulletin of the Society for the Study of Labour History*, no.40, Spring 1980, p.6. See also his: 'The Aristocracy of Labour Reconsidered', *7th International Economic History Conference* (Edinburgh, 1979).

14. Thorn, G., 'Statement Aristocrats and Sweated Labour: Rank-and-File Militancy in the London Bootmaking Union, 1885–1895', MA, Warwick, 1973; and the work in progress by A. Kershen on the Leeds clothing industry, 1850–1914, MPhil, Warwick, forthcoming.

15. See Matsumura, T., *The Labour Aristocracy Revisited: The Victorian Flint Glass-Makers, 1850–1880* (Manchester, 1983).

16. Bertrand Russell said that a pedant is someone who likes his statements to be true. However, this statement itself seems to be false! We do not in fact employ the term in that way. We reserve it to identify those who resist an unwelcome proposition, not by demonstrating that it is false, but by insisting that it is *unclear* by the standards of some new and demanding clarity. Characteristically, the pedant gives himself away by re-instating the terms or procedures against which he originally made his protest. This is exactly what Dr. Pelling did in his otherwise stimulating and important essay.

17. Musson, A.E., *British Trade Unions 1800–1875* (1972) tries to discount discontinuities. Others who acknowledge the reality of discontinuity attempt to account for it in other ways: through 'stabilisation' (Stedman Jones, *op.cit*) or through 'social control and liberalisation' (Foster, *op. cit.*) or perhaps 'deference' (Joyce, P., *Work, Society and Politics* (1980) pp.356).

18. Matsumura, *op. cit.*

19. *The Oxford Paperback Dictionary* (Oxford, 1979).

20. Braverman, H., *Labor and Monopoly Capital* (1974).

21. Turner, H.A., *Trade Union Growth, Structure and Policy* (1962),

pp.193–4, where the play between the two senses of 'skill' is clearly identified.

22. Samuel, R., 'The Workshop of the World: Steam Power and Hand Technology in Mid-Victorian Britain', *History Workshop Journal*, 3 (1977). Melling, J., 'Noncommissioned Officers: British Employers and their Supervisory Workers, 1880–1920', *Social History*, vol.5 (1980). More, *op. cit.*

23. See Thorn, *op. cit.*, and Kershen, *op. cit.*

24. Ure, A., *The Philosophy of Manufactures* (1835) p.368.

25. But see Zeitlin, J., 'The Emergence of Shop Steward Organisation and Job Control in the British Car Industry', *History Workshop Journal*, 10 (Autumn 1980).

26. See Anderson, P., 'The Origins of the Present Crisis', *New Left Review*, 23, and Thompson, E.P., 'The Peculiarities of the English' in his *The Poverty of Theory* (1978).

27. George, F.H., and Humphries, J.D., (eds), *The Robots Are Coming* (1974), p.164. I am indebted to Professor Howard Rosenbrock for this reference.

1. 'Our Class of Workmen': The Cabinet-Makers Revisited

David Blankenhorn

On 3 November 1884, at the fortnightly meeting of the Belfast Branch of the Friendly Society of Operative Cabinet-Makers, the men's anger exploded in a bitter resolution denouncing another group of workmen in their trade, the local branch of the Alliance Cabinet-Makers' Association. The Friendly Society accused the Alliance men not only of using 'vile threats' in an effort to 'extinguish all others not connected with their society', but also of using

> all means in their power to lower the position of the Trade in Belfast . . . While we . . . applaud the members of any kindred society that has for its object the Elevation of our Trade by using Legitimate and honest means to maintain our standing, we record our disapprobation of their conduct . . . in supplying the Employers in Belfast with their Furniture of Inferior order and make.[1]

The Belfast men were not alone. Throughout the 1870s and 1880s, this inter-union rivalry afflicted the two societies in cities and towns across the United Kingdom. Moreover, the conflicts reflected, not primarily mutual jealousies or jurisdictional disputes, but rather fundamental differences of organisation and purpose.

Despite the relative inattention of historians, these two unions represent the most important cabinet-makers' societies of the nineteenth century.[2] Moreover, they embody two distinct and competing strands of Victorian trade unionism. One strand, as seen in the FSOCM, represents what might be termed the trade unionism of threatened privilege. As the most 'respectable' craftsmen in an industry undergoing slow transformation, the élite cabinet-makers who comprised the Friendly Society struggled fiercely to

preserve their enclaves of relative security. Their battle was
waged on two fronts simultaneously: against their employers
above, who ceaselessly strove to reduce wages and alter
customary trade regulations; and against their fellow workmen below, who threatened to 'lower the position of the
Trade' by accepting reduced wages and producing 'Furniture
of Inferior order and make'.

The second strand of Victorian trade unionism, as seen in
the Alliance, represents a partial but significant realisation
of those principles which after 1888 came to be known as
'new unionism'. Alliance men were less 'respectable' than
their Friendly Society counterparts, possessing less training
and skill, and working primarily in shops which paid lower
wages and supplied the cheap furniture trade. The Alliance,
much more than the Friendly Society, recognised the
solution to their problems not in excluding the lower grades
of workmen but, instead, in embracing them: not as
preserving the status of the élite of the trade but, instead,
extending unionism throughout the trade. By the time of the
peak of the new unionist insurgency of 1891, the differences
between the new unionism of the Alliance and the craft
unionism of the FSOCM had become fully apparent: the
Alliance stood as the dominant union among the cabinet-
makers, having enrolled vast numbers of previously unorganised workmen from nearly every sector of the furniture
industry. By contrast, the older FSOCM remained virtually
stationary during the new unionist period, and entered the
1890s as an increasingly beleaguered preserve of the
industry's more privileged craftsmen.

Despite the richness of the surviving records for the
cabinet-makers, British furniture workers of the nineteenth
century have yet to find their historian. As a result, labour
historians of this period have failed to recognise in the
FSOCM the development of the New Model of mid-
Victorian unionism, and have consequently ignored the
importance of this development for the historiographical
controversy concerning a labour aristocracy during the
Victorian era. Indeed, G.D.H. Cole's dismissal of significant organisation among cabinet-makers, or A.E. Musson's
contention that the cabinet-makers were free of 'New

Model' tendencies, stand quite apart from the actual history of the trade.[3]

The evolution of unionism in the furniture trade parallels that of other similarly situated Victorian industries, particularly clothing and footwear, which also underwent decline and attendant disorganisation in the latter decades of the century. These analogous industrial transformations and trade union impulses played a significant — though not yet fully documented — role in both the new unionism of the late 1880s, and re-emergence of socialism in the trade union movement after 1880. So this case of two unions in a single industry perhaps adumbrates a more general mode of inquiry relevant to an important strand of the labour movement of the time.

'A SPIRIT OF COMBINATION'

'A spirit of combination has grown up among the working classes of which there has been no example in former times,' declared the *Poor Man's Guardian* in the autumn of 1833.[4] Those months of explosive activity by British working people have been explored by the Webbs, G.D.H. Cole and others,[5] but all of these historians have overlooked the formation in Liverpool, in August 1833, of 'the Trades Union, or Friendly Society, which has for its object the unity of all cabinet-makers in the three kingdoms'.[6] Yet unlike many other attempts that year at national unionism, including a Scottish cabinet-makers' union,[7] the Friendly Society of Operative Cabinet-Makers survived the collapse which followed the peak of 1833–4, and became, until surpassed in the 1880s by a younger and more vigorous society, the most important cabinet-makers' union in Britain.

The FSOCM was born amidst the turbulence of changing conditions in three broad areas: the cabinet trade itself, the national economy and the practice ('spirit of combination') of the day's trade unionism. Concerning changes in the cabinet trade, the first two decades of the nineteenth century saw cabinet-makers achieve a remarkable degree of organisation. Besides the powerful London Cabinet-Makers' Society, of which William Lovett for a time served as President, strong cabinet-makers' societies thrived in Birmingham, Edinburgh, Newcastle upon Tyne, Belfast and Dublin during this period. Yet despite this level of combination, the years between 1815 and 1830 were ones of decline for cabinet-makers. Apprenticeship restrictions were eroded in both law and practice, so that the supply of handicraftsmen gradually overtook the growing demand. More importantly, the end of the Napoleonic wars introduced a long period of chronic underemployment, most marked during the depression of the late 1820s, so that organised cabinetmakers consolidated their position in 'society' shops only at the price of swelling the numbers in the 'dishonourable' section of the trade.

But economic conditions changed in 1833, as workmen experienced the first year of good trade since the repeal of Combination Acts in 1824.[8] Trade societies quickly seized this moment of prosperity to form general combinations to demand better wages and enforce trade regulations. Led by the building operatives of Lancashire and the building and handicraft trades of the Midlands — areas which, along with Ireland, formed the original bases of FSOCM strength — the remarkable insurgencies of these months almost certainly influenced local cabinet-makers' leaders from these regions to meet in Liverpool in August to form a national union. Moreover, the building industry, the focus of this period's conflict, is the industry most related to cabinetmaking. Indeed, the report of the Liverpool 'General Trades Union' in August 1833 refers to 'Operatives in the Building Department', and the first FSOCM report refers to 'Societies in the House Furnishing department', so perhaps some formal connection existed initially between the two unions.[9]

Like the building workers, cabinet-makers in late 1833 and 1834 pursued a militant trade policy. In mid-1833 the Liverpool cabinet-makers struck and requested the recently formed Scottish National Union of Cabinet and Chair-Makers to 'acquaint those unaware of the strike in Liverpool'.[10] In 1834, the FSOCM executive received 'numerous claims . . . for strikes', many of which they were forced to disallow for financial reasons, despite the imposition of heavy strike levies throughout the year.[11]

Some additional evidence suggests that the trade militancy of the early FSOCM sprang, not only from the upturn in trade and the insurgency of the builders, but also from Owenite ideas of general unionism. For example, the Scottish National Union of Cabinet and Chair-Makers in 1833 appears to have been involved with attempts to form a general union of trades: either Robert Owen's Grand National Consolidated Trades Union, or the Owenite union, the Glasgow and West of Scotland Association for the Protection of Labour, founded by the Glasgow cabinet-maker Alexander Campbell in 1831. The Scottish cabinet-makers organised their union five months prior to the formation of the FSOCM. They certainly influenced the founders of the FSOCM, since the Scottish cabinet-makers were in frequent contact with Belfast, Manchester and Liverpool cabinet-makers throughout 1833 and 1834.[12]

Irish cabinet-makers who met in Londonderry in 1834 and 'came to the determination of joining the Trades Union, or Friendly Society' of cabinet-makers, were also possessed of the spirit of general unionism. One of their leaders, the Belfast cabinet-maker George Kerr, who also led the petition movement for the Dorcester convicts, clearly implies that 'the Trades Union' to which he belongs embraced numerous trades. That it indeed did appears confirmed by the fact that Kerr and four other cabinet-makers were promptly imprisoned by the mayor of Londonderry because of union activity.[13] Indeed, Irish cabinet-makers comprised fifteen of the forty FSOCM branches active in 1835. These links with the Irish and Scottish cabinet-makers, along with the likely connections with the builders' attempts at general union, clearly suggests that the Owenite spirit of general unionism that swept across the three kingdoms in 1833 and 1834 had also infected the working men who formed the FSOCM.

However, whatever militancy and fervor marked the FSOCM at its birth soon withered away during the crucible of depression which began in 1836. Entire branches collapsed or withdrew; membership fell rapidly. After 1836, the union was not to see 1000 members on its rolls again for thirty years. Membership during this period generally fluctuated between 400–600, although the statistician Dudley Baxter estimated in 1867 that approximately 39,000 cabinet-makers and upholsterers laboured in England and Wales alone.[14] The Liverpool FSOCM branch expressed its outrage in 1866 at 'the many thousand cabinet-makers in the United Kingdom' who work for wages 25–35 per cent lower than FSOCM standards.[15] Nevertheless, beleaguered though it may have been during these decades both by rapacious masters and 'dishonourable' workmen, the FSOCM, still based in Lancashire and the Midlands, remained Britain's largest union of cabinet-makers, comprised of the trade's best paid and most skilled craftsmen.

'A LEVEL WITH OTHER IMPORTANT TRADE SOCIETIES'

For the first thirty years of its history, the FSOCM operated as a loose federation of semi-autonomous local branches. The union's system of governance was simple. Every three years the membership elected a seat of management from among the larger branches to serve as a centre of communication and rudimentary record-keeping. Union finances, however, were decentralised: each year the seat of management computed the expenses of the various branches, divided total expenses by total membership, and charged each branch accordingly. Besides contributions and levies, entrance propositions — 7s 6d prior to 1874 — also generated union income. The union established a central fund in 1844 to insure against emergencies, but the fund remained quite small until the 1870s, when it reached several thousand pounds.

Prior to 1864, the union offered only four major benefits: allowances for travelling members or 'tramps' (computed by the mile), death benefits to widows (£4), insurance for tools (optional) and dispute benefits. After 1845 the union operated an optional sick fund, though it never became an important aspect of the union's early friendly benefit system.

But as the union grew larger and stronger during the prosperity of the 1860s and early 1870s, this simple system of government changed dramatically. The union's constitutional history from 1864 to 1874 reflects its transformation from a federation of local societies, each controlling its own funds, to a single, integrated association with a common purse. The union evolved an elaborate system of government which combined financial and administrative centralisation with democratic controls, and established an extensive network of friendly benefits supported by high contributions and entrance fees.

It was during the boom of the early 1860s that FSOCM leaders began to introduce sweeping changes in their union's constitution. The revised rules of 1864 mandated tool insurance of at least £10 for each member, as well as more generous funeral benefits. Three years later the union significantly enlarged its reserve fund, and required each member to contribute 3d per month towards its support.

The most important changes occurred in 1869. As the seat of management reported in 1870:

> Formerly each Branch controlled its own income and expenditure, and was debited a certain amount at the end of the year as a quota of expense, but under the present scheme all Branches are controlled by one uniform system.

Also in 1869 unemployment benefits were established and the sick fund extended to cover all members. For the first time, branch officers were paid a small salary, and a complex system of rules developed to prevent defalcations and insure efficient administration at every level of union government.[16]

In short, the FSOCM adopted during this decade the New Model of craft unionism that had been pioneered by the Amalgamated Society of Engineers in 1851 and — more importantly for cabinet-makers — by the Amalgamated Society of Carpenters and Joiners in 1861. In embracing the New Model, they established themselves, even more solidly than before, as a union comprised of the élite craftsmen of the trade.

By 1874 this transformation of the FSOCM into a New Model was complete. In October of that year, Alex Yule of the Manchester branch took office as the union's first full-time general secretary, at a salary of £2 per week. Also in 1874, superannuation and emigration benefits were added to the union's by now extensive network of friendly benefits.

Exclusiveness became both a central and necessary component of the New Model. Particularly in cabinet-making, which suffered from incomplete unionisation and competition from the 'cheap and nasty' trade, high contributions and entrance fees prevented most but the top craftsmen from joining; and even the better paid men often fell 'out of benefit' during times of high unemployment. But exclusiveness emerged as more than an indirect result of union policies, a passive attitude. The New Model FSOCM constitution also deliberately excluded undesirable applicants — either those considered incompetent or cheap workmen, or those likely to be an unprofitable drain on the friendly benefits system.

No historian has previously noted this constitutional development in the largest cabinet-makers' union of the period. Indeed, Cole states that cabinet-makers in the 1850s and 1860s were only organised in 'little local clubs', and Musson, in a recent historiographical essay on Victorian trade unionism, echoes Cole when he states that, in cabinet-making, 'little or no attempt was made to follow the New Model'.[17] This is incorrect. Evidence from the FSOCM rather supports the Webbs' older view — which Cole and Musson criticise — that many trade societies from 1852 through 1875 self-consciously adopted the spirit and central features of the ASE and ASCJ New Model. The FSOCM executive boasted in 1874 that the union's recent rule revisions placed it 'on a level with other important Trade Societies'.[18] Their self-assessment was almost, although not completely, correct: by 1874, the FSOCM had indeed adopted the major features of the two leading New Model unions. Particularly clear is the influence of the ASCJ — craftsmen in a closely related trade, to which the cabinet-makers often compared themselves in status and respectability.

Yet despite these developments, the FSOCM never quite placed themselves 'on a level with other important Trade Societies', such as the engineers, printers, flint glass-makers, shipbuilders and cotton spinners, which formed the most affluent ranks of the working class. Their size, fees, benefits and reserve fund remained lower than these other societies, reflecting the more significant reality that the cabinet-makers were less successful in controlling the market for their labour. The FSOCM resembled the ASCJ in more than constitutional development: like the builders, the cabinet-makers never became secure members of labour's 'aristocracy', but instead survived precariously on its periphery, as both industrial changes and competition from the unorganised and semi-skilled undermined their status as élite craftsmen.[19]

Of those historians who find the concept of a labour aristocracy a fruitful one in analysing mid-Victorian trade unionism, most agree that a weekly wage of between 30–5 shillings marks the dividing line between the 10–20 per cent best-paid working men and the rest of the labouring population.[20] The wage data from Table 1.1 suggest that the cabinet-makers of the FSOCM were borderline aristocrats: very few earned as much as 30 shillings before 1865, but most branches had crossed this threshold by the early 1870s. Yet few branches outside Manchester earned as much as 35 shillings as late as 1875, and many smaller branches remained well below even the 30 shillings level. Most fluctuated near the border which separated relative comfort and respectability from poverty. Moreover, the mid-1870s constituted the FSOCM's moment of greatest strength: almost every branch suffered wage reductions during the depression of the late 1870s and early 1880s, and faced growing challenges from semi-skilled and non-society workmen.

This characterisation of the FSOCM is supported by comparisons of their wages with those of other trades. During the mid-Victorian years FSOCM earnings stand roughly equivalent to those of upholsterers, organised building trades workers, and also to those of servitors, or middle-grade flint glass-makers, each group occupying an

Table 1.1: FSOCM Standard rates of wages and hours of labour for 25 selected branches, at two-year intervals, 1865–1875

Branch	1865 wages	1865 hours	1867 wages	1867 hours	1869 wages	1869 hours	1871 wages	1871 hours	1873 wages	1873 hours	1875 wages	1875 hours
Manchester	30	56½	32	56½	32	56½	32	56½	34	54	38	54
Birmingham	26–8	59½	28	59½	28	59½	28–30	59½	30–2	54	32	54
Chester	28	59	30;5–25% off L.B.	59	30;5–25 off L.B.	59	30;15 L.B.	54	30;5 off L.B.	54	32;7½ 7½ on L.B.	54
Derby	26	58½	26	59	26–30	59	26–30	59	28–30	54	32	54
Wolverhampton	26	60½	26–8	60	26–8	60	28–30	56½	28–30	56½	28–30	56½
Rochdale	28	56½	28	56½	28	56½	30	56½	32	54	34	54
Shrewsbury	24	59	—	—	—	—	25	59	25	56½	26–7	56½
Sheffield	29	58	29	56½	30	56½	30	51	32	51	36	51
Preston	26	56½	28	56½	28	56½	28	56½	30	54	34	54
York	23–4	58	25–6	58	25–6	58	25–6	58	25–6	54	26–8	54
Liverpool	30	59	30–2	59	30–2	59	30–2	59	32	54	34	54
Halifax	26	57	28	57	28	57	28	54	30	54	31–2	54
Dungannon	22–4	60	22–4	60	22–4	60	22–4	60	26	60	26	60
Huddersfield	26	58½	27	58½	28	56	28	54	31	54	33	54
Newcastle-under-Lyme	L.B. 15–20% off	55½ 56	L.B. 15–20 off	56½	L.B. 15–20 off	56½	L.B. 10–20 off	56	30	54	31–6	54
Keighly	24	57½	26	56	26	56	26	56	26	54	27	54
Londonderry	—	—	20	60	22	60	—	—	24–6	57	24–6	57
Nottingham	25	58½	27	58½	27	58½	27	58½	30	54	31–6	54
Hereford	20–4	58–62	22–4	59	22–4	59	22–4	59	22–4	54	25–7	54
Bolton	28	58½	28	56½	30	56½	30	54	32	56	36	54
Blackburn	26–8	57½	28	55	28	55	30	54	32	54	36	54
Belfast	22–4	59	22–8	60	22–8	60	22–8	60	30	54	32	54
Bradford	21–6	59	23–7	56½	26–9	56½	26–9	54	29–31	54	31–3	54
Southport	28	57½	30	57½	30	57½	30	54	34	54	38	54
Lancaster	24	60	26	60	26	60	26	60	—	—	30	54

Source: FSOCM, *YR*, 1865, 1867, 1869, 1871, 1873, 1875.

analogous borderline position in the labouring hierarchy. The magnitude of the cabinet-makers' wage increases in the 1860s and 1870s, as compared to other trades, points to the same conclusions: between 1850 and the 1880s, the already highly paid trades such as printing and engineering experienced relatively modest increases, while the FSOCM were more like both the flint glass servitors and building workers, who enjoyed increases of over 40 per cent.[21] In short, the evidence on levels of earning — which Hobsbawm considers the primary criterion for determining labour aristocratic status[22] — further suggests that FSOCM members did indeed form a contingent of British labour's aristocracy, but also that their aristocratic status was a precarious one.

'. . . FIT TIME TO BE UP AND DOING'

An important flaw in the Webbs' analysis of mid-Victorian New Model unions is the view that these societies pursued timid, even class collaborationist, trade policies.[23] As several subsequent historians have shown, the third quarter of the century, particularly the 1860s and early 1870s, constitutes a period of considerable trade union militancy, when unionists by no means accepted uncritically the canons of middle-class political economy.[24] The FSOCM executive, reporting an expensive and unsuccessful strike in 1859, noted that

> In these movements there is something to be regretted; but what can we do? Labour has no other resource to fall back upon in times of oppressions . . . It has been astounding to observe that men of rank, and whose philanthropy is of great pretension, denounce these movements on the part of the working men, and at the same time withhold every other means of redress.[25]

Branch movements to obtain higher wages and better working rules, or conversely to resist encroachments by employers, reveal the most dramatic expression of union trade policy: at least 46 significant strikes — involving ten or more workmen — occurred in FSOCM workshops during the 1845–1875 period. Yet even more illuminating are the

small-scale workshop conflicts over the maintenance of
standards and preservation of trade customs, usually involving only one or a few men, that punctuate the history of each
branch. For several reasons, FSOCM workshops were
particularly marked by this type of localised conflict.

First, even in many 'respectable' workshops, unionisation
remained incomplete, and it is a commonplace of industrial
relations that conflicts are more numerous in skilled trades
which are incompletely unionised. Moreover, this general
problem was radically accentuated by the prevalence of the
piecework in the trade. The issue is exceedingly complex.
The Webbs, attempting to explain the persistence of
payment by result despite the declining use of the London
Books as standards, argue that

> In consequence of the great changes in and multiplication of patterns,
> and the alteration of processes, the lists have long since been obsolete,
> and no one has yet found it possible to classify the innumerable jobs
> now involved in the manufacture of furniture . . . So strong, however,
> has been the tradition and custom of piecework in the trade that none
> of the various unions which have from time to time arisen during the
> last half century have been able to stand out for time wages.[26]

The Webbs' analysis is very misleading in several respects.
Although the multiplicity of patterns certainly made classification complex, this fact alone fails to explain the
increasing 'obsolescence' of the standard books. The Belfast
cabinet-makers admitted as early as 1822 that the 'constant
change of Fashion in the style of Cabinet-work renders it
difficult, if not impracticable . . . to embrace particularly the
price of every article', yet insisted that the price book
contain 'the general principles for ascertaining the value of
Work'.[27] But this method of payment requires strong
unionisation to protect its integrity, and this essential
element was progressively weakened during the Victorian
era. The standard books of prices fell into obsolescence for
social, not technical, reasons. In better organised trades,
such as cotton spinning, unions successfully enforced exceedingly complex price lists. Yet in the cabinet trade,
employers manipulated the piecework system to batter
down wages, by reclassifying jobs or abandoning standard

price lists altogether in favour of 'lump' work — the men's epithet for individual bargaining which eroded standards.

The Webbs were also mistaken in saying that a 'strong . . . tradition and custom of piece-work in the trade' prevented cabinet-makers from seeking or obtaining time wages. While book prices remained the practice in a few of the most expensive shops, most FSOCM branches, aware of the system's dangers in the absence of strong unionisation, demanded and generally obtained day wages. At least as early as the 1860s the FSOCM executive required new branches to 'conform to our laws by adopting day-work when opportunity occurs';[28] and the Belfast FSOCM in 1874 considered it 'the duty of every Branch to do away with piecework altogether' since it was 'quite antagonistic to the spirit of the union'.[29] Bastardised piecework or 'lump' work existed in some shops, not because of irrational 'tradition and custom', but rather because employers imposed it and the local union was too weak to resist it. In all cases, struggles over the methods of payment insured continual conflict between cabinet-makers determined to preserve their craft autonomy and employers strong enough to threaten it.

Both branch movements and localised workshop conflicts were directed against employers. But the third modality of FSOCM trade policy — regulation of the market for labour — was directed, not at employers, but at other workmen in the trade. The FSOCM executive in 1871 explained this dichotomous policy in which

> a great portion of our funds are absorbed protecting our labour from the rapacity of employers who want to get rich too soon, and from the machinations of unprincipled journeymen, who degrade themselves in the eyes of their fellow workmen by submitting to be made the pliable tools of despotism and tyranny.[30]

The primary analytic value of the concept of a labour aristocracy lies precisely in its denotation of this complex triangular relationship among skilled craftsmen, their employers from above and their less privileged fellow workmen from below. Harriet Martineau penetrated the essence of that ubiquitous Victorian word, 'respectable' when she noted that 'All labour for which there is a fair demand is respectable'.[31] The 'aristocracy' were 'respectable' because they insured a 'fair demand' for their labour by restricting its supply. One of the first working men to use the term 'aristocracy of labour' was Samuel Jacobs, the secretary of the Bristol FSOCM branch during the mid-1840s. Jacobs urged in 1845 that the Peoples' Charter be pursued through a general union of trades embracing the 'aristocracy of labour', by which he meant those workmen who 'maintained the price for their labour'.[32]

The FSOCM tried in a number of ways to control the market for cabinet-makers. High entrance fees and contributions served as barriers against undesirable members. In addition, many men who applied for admission were rejected by branch leaders — particularly the unapprenticed, those accepting less than the standard rate, or those likely to draw heavily upon the benefits system.

Concurrently, the union fought a continuous battle to restrict the number of apprentices and boy labourers allowed to work in each shop. The Victorian period saw the serious erosion of apprenticeship regulations in nearly every skilled trade, and cabinet-making was no exception. But it would be misleading to generalise from the experience of London craft societies, as some historians seem to have done. Even as late as 1892, a system of apprenticeship in the cabinet trade generally prevailed in the areas outside London, especially in the smaller towns.[33] By allowing each branch to form its own by-laws concerning apprentices, the FSOCM recognised that its ability to restrict entry into the trade was declining; yet at the same time the system maximised the possibilities for continued control.

Cabinet-makers in the FSOCM also fought relentlessly for closed shops in the 'respectable' segment of the trade. Unable to dominate the trade as a whole, the FSOCM sought to preserve for themselves a secure enclave within it. Each branch kept a list of *'bona fide* shops' — shops that paid the standard rate — and prohibited members from seeking work in 'dishonourable' shops. Whenever possible, the men prevented non-society men from obtaining employment in FSOCM dominated shops, and large branches such as Manchester and Liverpool achieved considerable success in controlling 'respectable' establishments.

The 1860s and early 1870s represent the peak of the FSOCM's strength and prestige in the nineteenth century. These years were 'a fit time to be up and doing', as the secretary of the militant Liverpool branch put it.[34] Yet ironically, their strongest moment also revealed their basic weakness. The union never achieved full control over even its own sphere of the trade, and the onset of depression in the mid-1870s only accentuated this underlying problem. More fundamentally, the Great Depression was to throw into bold relief the long-term structural transformations in the furniture industry itself. These industrial developments were creating new conditions in which cabinet-makers' societies had to exist: conditions which frustrated the growth of the FSOCM, and promoted the emergence of another, differently organised society, the Alliance Cabinet-Makers' Association.

'MEN WHO CALL THEMSELVES CABINET-MAKERS'

Cabinet-making, like textiles, experienced an industrial revolution based on steam power in the early nineteenth century. Yet unlike the case of textiles, it was not a revolution that transformed the basic division of labour within the industry. The major processes of furniture production which were first mechanised — the sawing and planing of the planks, and the cutting of veneers — soon came to be undertaken by specialist contractors working in saw mills, and left virtually intact the basic work of the skilled cabinet-maker, who continued to produce furniture almost entirely in the traditional handicraft method. Further sophistication of woodworking machinery did not occur until the 1850s and 1860s, when the band-saw, which cut wood with an endless band or ribbon-saw, and fret-cutting machines for tracery, fretwork and other internal cutting, achieved widespread adoption. Yet initially these machines did not so much replace the handicraft labour of the

cabinet-maker as make it more productive without destroying its skill. The machines' impact during the 1850s and
1860s on the central processes of furniture making appear
limited and indirect, and their major effect was to cheapen
production.[35]

The period after 1875, however, saw an unprecedented
extension of woodworking machines, especially in areas
outside London, and it is during this period that the
cabinet-makers' trade societies first begin to complain that
machines were depressing their wages and threatening their
status as skilled craftsmen.

By the 1880s, skilled cabinet-makers almost everywhere
saw '[m]achine-power . . . taking the place of handicraft to
an extraordinary extent'. Fully mechanised 'cabinet factories' sprang up in a number of towns, provoking angry
FSOCM leaders to complain that 'the hundreds of men that
work in factories and call themselves cabinet-makers could
neither set out a job or lay a veneer'.[36]

The new machines radically accentuated the specialisation
of labour, especially since '[w]here cabinet-making by
machinery is adopted as a system, each firm generally makes
one kind of work as a speciality'.[37] Machines also, as one
cabinet-maker put it in 1892, enable capitalists to 'employ
boys to do men's work'.[38] In short, machines discounted the
skills of the 'practical workman', and created new grades of
workers to correspond to a changed division of labour.

However, much more than increasing mechanisation
imperiled the status of the cabinet-maker. Traditional
Victorian historiography teaches that the evolution of the
Victorian economy was 'centripetal', producing an increasingly centralised mode of production and an increasingly homogeneous working class. But nineteenth-century
cabinet-making, like clothing and footwear, underwent
'centrifugal' evolution: its adaptation to market and technical conditions promoted the radical decentralisation of
production and the fragmentation of the industry through
subcontracting.[39] These tendencies toward 'industrial disorganisation' developed concurrently with the trend towards
mechanisation, and in many cities, particularly in London,
represented the dominant logic of the industry's evolution

through the 1880s. Though the impact of machinery was important, it none the less remained limited and uneven. Moreover, the early and mid-century woodworking machines which cheapened raw materials actually stimulated the growth of small masters, subcontracting and sweating.

The roots of the disorganisation of the cabinet trade reach back to the earliest years of the Victorian period. The 'dishonourable' trade spread rapidly in London in the 1830s and particularly in the depressed years of the 1840s, as middlemen, seeking a supply of cheap furniture, established 'linen drapers' or large furniture warehouses and subcontracted work to poor 'garret-masters' from East London. A typical small master, himself under contract to wholesalers, subcontracted all of his sawing, turning, fretcutting, polishing, upholstery and carving to outsiders and, as Booth described it, '[w]e find men who call themselves cabinet-makers making only one, or maybe two or three articles, and this fact, coupled with that of a small system of production, may be said to be the leading characteristic of the whole district'.[40] The decentralisation of the trade proceeded further in London than in other areas, but conditions prevailing in the East End during the 1870s and 1880s were replicated on a smaller scale in provincial towns.

Thus both machinery and industrial disorganisation, though separate, produced similar problems for the skilled craftsman. As the Alliance Cabinet Makers' Association saw it in 1881, both the 'sub-division of labour, which diminishes the necessity of skill in certain productions, and the employment of machinery . . . are directed to the further subjugation of the workman'.[41] Both trends threatened the 'subjugation of the workman' and 'diminished' traditional craft skills because they created new grades of workmen to compete with traditional craftsmen: those who were un-apprenticed and semi-skilled; those who operated wood-working machines; those who made cheap furniture for low wages; those who were unorganised and weak in relation to their employers. In short, they included men who 'called themselves cabinet-makers', yet lacked the artisanal status and market power of élite craftsmen.

The 1860s marked a turning point in the London cabinet

trade. As disorganisation in the furniture industry accelerated, East London superseded the West End as the centre of the industry. The 1860s also saw the birth of a new cabinet-makers' society, based initially in East and North London. The new Alliance Cabinet-Makers embraced grades of workmen who were ineligible for the more exclusive FSOCM branches. In the 1870s, the young society expanded into the provinces as well, enrolling the lesser paid, lesser skilled men who were rapidly growing in number in response to the cumulative transformations of the industry.

'THE WHOLE CABINET TRADE'

'The upheaving of labour is almost universal throughout the kingdom,' declared the *Bee-Hive* in November 1865.[42] Like the FSOCM over thirty years earlier, the Alliance Cabinet-Makers' Association was born during a major upsurge of trade union militancy. The labour insurgency of this period is remarkable for its quality as well as its breadth, as lesser skilled and lower paid workers began to organise themselves to press their demands upon employers.

In October 1865, only two cabinet-makers' societies existed in London: the presitious West End Society with about 300 members, and 'a smaller one in the East-end of the town . . . the member of the latter being less than fifty, and they were colloquially known among some of the employers as the forty thieves'. As a London cabinet-maker later recalled, the Alliance began during that month

> mainly with a view to organise the trade in the East End, which undoubtedly had been neglected by the more prosperous body of men in the West End, who claimed the exclusive right to work in West End shops, as also did the Manchester Union (FSOCM) in the best shops in the provinces.[43]

On 11 October, a meeting of London cabinet-makers 'from the principal shops in the eastern and northern districts' determined to organise a 10 per cent wage movement among the city's non-society cabinet-makers.[44]

Initial efforts met with only partial success, so on 22 November over 1000 cabinet-makers and shop delegates agreed by acclamation to 'the formation of an association' in order to 'obtain the advance of ten per cent throughout the whole trade'.[45] Nearly 800 men agreed that night to join the new society, and by the year's end the 10 per cent movement was successfully completed.

Yet even in the flush of victory, the position of the fledgling society remained precarious, especially as the boom ended and a depression emerged during 1866. The men's basic weakness is illustrated by the fact that, unlike the West End society or the FSOCM, Alliance branches were unable to enforce book prices or indeed any general standard of payment for their labour. Instead, the Alliance were forced to imitate the practice of the East London 'forty thieves': each individual shop drew up its own price list, and attempted to enforce these prices as minimum standards. Employers frequently used this 'shop bargain' system of payment systematically to batter down prices.

The two great periods of expansion and transformation for the Alliance were the early 1870s and the late 1880s. The analogy between these two periods is strong. The large changes in the early 1870s — stilled by the onset of the Great Depression — foreshadowed the even larger changes of the late 1880s. In both periods, the union grew spectacularly and asserted itself to improve the wages and working conditions of its members. More fundamentally, the Alliance in both periods expanded its social as well as geographical base, and reached out, far more than did rival cabinet-makers' societies, to the lesser paid and workmen of the trade.

The first period of growth began in May 1872, when the Alliance amalgamated with the East London Cabinet-Makers' Society. The Alliance executive called the amalgamation with the 'forty thieves' the 'most successful step that has ever been taken by the members, not only to benefit our own individual interests, but to promote the prosperity of the whole Cabinet trade'.[46] This amalgamation signalled a broader upsurge; membership doubled during the first six months of 1872, reaching 491 by July.

Another successful wage movement later that year launched the society into a three-year period of phenomenal growth. The Alliance for the first time extended beyond London, as branches opened in Hastings, Reading, Bath and Manchester. Progress became even more dramatic in 1874, as 705 members and 12 branches joined the union, and at least five branches won wage increases. By 1875 the Alliance boasted 28 branches — seven in London — and nearly 2000 members. A number of the provincial branches were organised in the South and West of England, but even more were opened in the northern strongholds of the FSOCM: Manchester, Liverpool, Sheffield, Oldham, Bury, Bolton, Burnley and Bradford.

Numerous amalgamations swelled Alliance membership rolls during these months. Particularly illustrative is the amalgamation with the Manchester Amalgamated Society of Cabinet and Chair-Makers in late 1873. In August of that year, the Manchester Branch of the FSOCM voted to donate £10 to the city's 'Second Body of Cabinet-Makers' who were

out on strike for an advance of prices. Their funds are nearly exhausted . . . The members of that Society work in a very inferior class of shops, and are principally engaged in manufacturing goods for brokers.[47]

This 'Second Body' must certainly have been the Amalgamated Society. By early 1874, Alliance reports show that these workmen from 'a very inferior class of shops' had both amalgamated with the Alliance and won their strike 'for an advance of prices'.[48]

Though the Alliance successfully amalgamated with 'inferior' societies, their overtures to the FSOCM in 1873 were firmly rebuffed. Yet by 1876 the FSOCM had completed an amalgamation with the prestigious West End Cabinet-Makers' Society as 'an absolute necessity for the better protection of all those engaged in the manufacture of first-class furniture'.[49] The West End society fully agreed with the FSOCM's rejection of 'Alliance principles', since Alliance members 'have no minimum wage, nor fixed price, they can accommodate any employer, as far as price goes. Whereas our members are compelled to act on principle.'[50]

Yet despite the disdain of the more prestigious societies,

the Alliance grew in just ten years, from 1865 to 1875, from a meeting of London's poorer cabinet-makers to one of the two largest cabinet-makers' unions in Britain. In 1875 membership stood at nearly 2000, only 150 less than the FSOCM. Of course, these were years of expansion for most British trade unions. But the rapid emergence of the Alliance is particularly significant since its growth derived in large part from its policy of embracing the lower grades of workmen who were not qualified or willing to join the FSOCM or West End Society. The years 1865–1875 saw the FSOCM reshape its constitution along the lines of the New Model: its amalgamation with the West End Society in 1876 is both the symbol and culmination of this development. The Alliance moved in precisely the opposite direction. The lean years of the Great Depression were to obscure and delay these trends, so that the implications of the two unions' increasingly divergent trajectories would not become fully apparent until the late 1880s. But the roots of these differences were firmly laid in the 1860s and early 1870s.

'OUR CLASS OF WORKMEN'

'[W]e have practically debarred any other society being organised which could . . . reach our class of workmen,' boasted the FSOCM executive in 1877, soon after it had rejected a working agreement proposed by the Alliance.[51] This self-congratulatory statement was not wholly true, for some evidence suggests that a number of highly skilled cabinet-makers were attracted by the Alliance's lower fees and contributions; nevertheless it accurately conveys the basis of the FSOCM's relationship to the Alliance during these years.[52]

The Alliance did not record wage rates prior to 1881 and after that only classified each branch under one of three wage 'categories' for the purpose of computing friendly benefit and contribution levels. Yet despite these imprecisions, available wage data reveal that nowhere did Alliance men receive higher wages than their FSOCM counterparts, while in several areas — at least London, Bradford, Preston in 1882, Manchester in 1890 — FSOCM wages were higher even than the broad categories used by the Alliance.[53]

But more important is the reason why the Alliance could not record the precise wages of its members: the Alliance, unlike the FSOCM, was too weak to enforce branch-wide wage standards. While FSOCM rules required that 'When a member obtains employment in a branch, he must see the secretary to ascertain the standard wages',[54] Alliance rules merely forbade members from accepting a piece of work 'at a lower price than that fixed by the members working in the same shop'.[55] This fundamental difference was a source of bitter animosity toward the Alliance from the FSOCM, who accused the Alliance of under-working those craftsmen who 'are compelled to act on principle'.

A comparison of the two unions' contributions and entrance fees shows that the Alliance weekly contributions stood at 7½d in 1875, compared to 9d for the FSOCM. The Alliance admitted new members for a flat 4 shillings, while the FSOCM charged entrance fees ranging from 7s 6d to several pounds, depending upon the age and desirability of the applicant. These differences widened throughout the 1880s. In 1880, the FSOCM raised its weekly contributions to 1 shilling, while the Alliance's contributions had been lowered to only 5d. In 1881 the Alliance introduced a graduated scale of contributions and benefits which lowered entrance fees and allowed workmen to join the Alliance for as little as 4d per week. Finally in 1887, the Alliance initiated a Trade or Partial Benefit Section, in which low-paid workmen received strike pay, tool insurance and legal assistance for only 3–4d per week.

The two unions' levels and priorities of expenditure point to the same conclusions: overall, the FSOCM spent more money on a wider network of mandatory friendly benefits than did the Alliance. However, the Alliance spent more than the FSOCM on strike pay in 14 of the 18 years from 1875 to 1892, and in every year after 1877 except one. So despite the fact that Alliance rules stipulated 'shop bargains' instead of branch-wide standards, the Alliance allocated significantly more money than did the FSOCM to strike support, both in absolute terms and, even more, as a proportion of total expenditure. During the early and mid-1880s especially, the FSOCM adopted increasingly quiescent trade policies, as the executive used its influence to discourage strikes in an effort to conserve diminishing funds. In fact, the conservatism in trade matters which is often ascribed to New Model unions during the third quarter of the century accurately describes the FSOCM not during the 1860s and early 1870s, but instead during the lean years which followed. Of course, the Alliance too suffered from declining memberships and strained funds during these years, and used dispute benefits more to resist reductions than to demand increases, but its position in the trade and its resulting attitude was altogether more aggressive.

The rate of membership turnover in each union reveals both its ability to recruit new members and, more importantly, the 'class' of workmen it embraced. Arrears accounted for the vast majority of exclusions in both societies. But despite the Alliance's lower fees and contributions, far more Alliance than FSOCM members fell 'out of benefit' each year through non-payment. This higher turnover and less stable membership clearly suggests that the Alliance enrolled a poorer and less secure stratum of workmen, more vulnerable to the vicissitudes of the trade cycle and generally more difficult to organise on a permanent basis. The figures on exclusions also show that the Alliance was significantly more energetic than the FSOCM in recruiting new members. Particularly revealing are the data for 1891: the Alliance lost 1357 members through non-payment, yet experienced an overall increase of 1082, while the FSOCM excluded only 134 members and grew overall by a meagre 24 members.[56]

The record of Alliance branch openings and amalgamations in this period further demonstrates the union's policy of expansionism, rooted in its commitment to organising cabinet-makers who were ineligible for the FSOCM. The Alliance's Birmingham branch forms an excellent example. For many years Birmingham had been a stronghold of the FSOCM. In 1876 the FSOCM Birmingham branch boasted 150 members, which, as the branch's secretary recalled to Sidney Webb in the early 1890s, 'was practically all the skilled men in the town'. Yet, as an Alliance man explained to Webb, the FSOCM had long been

> a very conservative body, admitting only those employed in shops making really first-class work, with the result that as the number of shops making cheap goods increased, so the number of non-unionists increased also.

In 1876 'these men making cheap goods succeeded in organising themselves into the Alliance'. The FSOCM officer whom Webb interviewed concurred that this 'large class of men called "Sloggers", making common deal or goods' had been 'always refused admission to the [Friendly] Society, and remained unorganised until 1876 when they formed the branch of the Alliance.'[57] There are many other examples; indeed, in 1890 the Alliance executive hoped that 'at no distant date . . . means may be devised for one gigantic organisation of the whole Cabinet Trade in the United Kingdom'.[58]

An important way in which the Alliance sought to establish 'one gigantic organisation' was widening its membership qualifications to include a wider range of occupations within the furniture industry. FSOCM rules allowed the admission of cabinet-makers, chair-makers, carvers and turners, yet the union remained almost wholly comprised of cabinet-makers throughout its history. The Alliance, by contrast, organised entire branches of chair-makers, fancy cabinet-makers, turners, carvers, and cabinet case-makers during the depression years. Shop-fitters were admitted to the Alliance as early as 1875. This trend greatly accelerated during the new unionist insurgency. By 1891 woodworking machinists, coach finishers, mill sawyers, fret-cutters and upholsterers were welcomed into the Alliance. Several years later, French polishers were added to the list of eligible occupations. Whether the Alliance was enrolling 'sloggers' in Birmingham and the East End, or woodworking machininists in London and East Manchester, the union consistently sought out precisely those new classes of workmen which were being created by the structural developments of mechanisation and disorganisation within the industry.

Each of these tendencies within the Alliance dramatically escalated and came to fruition after 1887, as the revival of trade allowed the Alliance to reap the advantages inherent in its policies. The membership figures of Britain's three national cabinet-makers' unions speak for themselves. Between 1887 and 1891, the FSOCM grew by 22 per cent, the Scottish union of cabinet-makers by 76 per cent, and the Alliance by 387 per cent, becoming by far the largest and most important trade union in the industry.[59]

The Alliance entered the 1890s as the new union of the trade: an organisation embracing the lesser paid and previously unorganised and, through its increasing heterogeneity, beginning to evolve from a craft toward an industrial basis of organisation. The FSOCM, which in its moment of birth nearly sixty years previously had sought 'the unity of all cabinet-makers in the three kingdoms', had finally been overtaken by a more vigorous society, and entered the 1890s as an overshadowed enclave of craft tradition increasingly out of harmony with a changing industry.

NOTES

All works printed in London unless otherwise noted.

1. FSOCM Belfast branch minute-book, 3 November 1884.
2. J.L. Oliver's *The Development and Structure of the Furniture Industry* makes almost no mention of workers. By far the best study of the London furniture industry is P.G. Hall's *The Industries of London Since 1861* (1962), yet Hall concentrates much more on the unorganised workers of the East End than on trade unionism in the industry. The analysis of the furniture industry in Gareth Stedman Jones's *Outcast London* (1971) derives primarily from Hall's work. S. Higgenbottam, *Our Society's History* (Manchester, 1939), and J. Connolly, *The Woodworkers 1860–1960* (1960), chronicle the history of the Amalgamated Society of Woodworkers (with whom, through amalgamation, the Friendly Society of Operative Cabinet-Makers merged in the twentieth century), but make only the briefest references to the Friendly Society or the Alliance Cabinet-Makers. N. Robertson, 'A Study of the Development of Labour Relations in the British Furniture Trade', B.Litt, Oxford, 1955, is concerned primarily with the twentieth-century furniture industry; moreover,

his cursory discussions of the Alliance and FSOCM in the nineteenth century are riddled with serious factual errors.

3. G.D.H. Cole, 'Some Notes on British Trade Unionism in the Third Quarter of the 19th century', *International Review for Social History*, II (1937), reprinted in E.M. Carus-Wilson (ed.), *Essays in Economic History* vol.III (1962), p.205; A.E. Musson, *British Trade Unions 1800–1875* (1972), p.50.
4. *Poor Man's Guardian*, 19 October 1833.
5. S. and B. Webb, *History of Trade Unionism* (1911); G.D.H. Cole, *Attempts at General Union: a study in British Trade Union History, 1818–1834* (1953).
6. G. Kerr, *Kerr's Exposition of Legislative Tyranny and Defence of the Trades Union* (Belfast, 1834; p.5. *Haliday Pamphlets*, 1611 (3), Royal Irish Academy.
7. See the minute-books and financial records from 1833–1837 of the Edinburgh branch of the Scottish National Union of Cabinet and Chair-Makers, National Library of Scotland.
8. Musson, *op. cit.*, p.40. 'Minutes of Evidence taken before the Select Committee appointed to Enquire into the Present State of Manufactures, Commerce and Shipping', published in *Edinburgh Review*, October 1833.
9. *Liverpool Chronicle*, 17 August 1833; FSOCM, *YR*, 1834–5.
10. Minute-book of the Edinburgh branch of the Scottish National Union of Cabinet and Chair-Makers, 28 May 1833.
11. FSOCM, *YR*, 1834–5.
12. I. MacDougall, 'The Edinburgh branch of the Scottish National Union of Cabinet and Chair-Makers, 1833–1837', *The Book of The Old Edinburgh Club*, XXXIII, part 7 (1969) pp.2–10. This article is useful and suggestive, but incomplete or incorrect in its reference to the FSOCM (or what is misnamed the 'English National Union of Cabinet-Makers'); minute-books for 1833 and 1834 of the Edinburgh Branch of the Scottish National Union of Cabinet and Chair-Makers.
13. Kerr, *op. cit.*, pp.5–16.
14. R.D. Baxter, *The National Income of the United Kingdom* (1868) p.88, cited in J. Burnett, *Useful Toil* (1977 edn) p.261.
15. *Bee-Hive*, 3 March 1866.
16. FSOCM, 1874 rule-book.
17. Cole, *op. cit.* (1937), p.205; Musson, *op. cit.*, p.50.
18. FSOCM, 1874 rule-book.
19. See H. Pelling, *Popular Politics and Society in Late Victorian Britain*, (1968), pp.51–2.
20. E.P. Thompson, *The Making of the English Working Class* (1968) p.238; Burnett, *op. cit.*, p.253; see also E.J. Hobsbawm, 'The Labour Aristocracy in 19th century Britain', in his *Labouring Men* (1964), pp.278–84.
21. Burnett, *op. cit.*, pp.261, 270; K. Katanka and E. Frow *1868: Year of the Unions* (1968), p.99; T. Matsumura, 'The Flint Glass-Makers in the Classic Age of the Labour Aristocracy', PhD, Warwick, 1976,

pp.80–93.

22. Hobsbawm, *op. cit.*, p.243. For criticism of this view, see Pelling, *op. cit.*, ch.3, esp pp.52–6.

23. 'Whether it is the Masons or the Tailors, the Ironfounders or the Carpenters, we see the same abandonment by the Central Executive of any dominant principle of trade policy, the same absence of initiative in trade movements, and the same more or less persistent struggle to check the trade activity of the branches.' Webbs, *op. cit.*, p.304.

24. Cole, *op. cit.* (1937), p.219; R.V. Clements, 'British Trade Unions and Popular Political Economy, 1850–1875', *Economic History Review*, 2nd ser., XIV (1969), pp.93–104; R. Harrison, *Before the Socialists*, (1965), pp.6–18.

25. FSOCM, *YR*, 1859.

26. S. and B. Webb, *Industrial Democracy*, (1902 edn), pp.301–2.

27. Belfast Cabinet-Makers Piece Price List (1822), ms. and printed versions in Modern Records Centre, Warwick University.

28. FSOCM, *YR*, 1867.

29. FSOCM, Belfast branch minute-book, 26 April 1874; 12 November 1877; 3 February 1879; 7 April 1884.

30. FSOCM, *YR*, 1871.

31. H. Martineau (cited in lecture; source unknown).

32. *Lloyds' Weekly Newspaper*, 4 May 1845. I am grateful to Mike Shepherd for this reference.

33. *Royal Commission on Labour*, *Minutes of Evidence*, Group B, Parliamentary Papers, 1893–4, XXXIII, q.19, 798.

34. *Bee-Hive*, 3 March 1866.

35. R. Samuel, 'Workshop of the World: Steam Power and Hand Technology in Mid-Victorian Britain', *History Workshop*, III, (Spring 1977), pp.37–8.

36. FSOCM, *YR*, 1888.

37. *Cabinet-Maker and Art Furnisher*, 1 September 1881.

38. *Royal Commission on Labour*, *op. cit.*, q.19, 926.

39. J.A. Schmiechen, 'Sweated Industries and Sweated Labour: A Study of Industrial Disorganization and Worker Attitudes in the London Clothing Trades, 1867–1909', PhD thesis, University of Illinois, Champaign-Urbana, 1975, pp.1–2; E.J. Hobsbawm, *Industry and Empire* (1968), p.71.

40. C. Booth (ed.), *Life and Labour of the People of London*, p.319; see also *Lords Committee on Sweating*, p.p.1888, XX, q.2207.

41. Alliance Cabinet-Makers' Association, recruitment brochure, 1881, WTUC, B, 60.

42. *Bee-Hive*, 18 November 1865. France and a number of other countries also experienced 'labour revivals' in the early 1860s. See H. Collins and C. Abramsky, *Karl Marx and the British Labour Movement* (1965), p.14.

43. 'A Chat About Trade Societies', *Cabinet-Maker and Art Furnisher*, 1 May 1890.

44. *Bee-Hive* 21 October 1865.
45. *Bee-Hive* 25 November 1865.
46. ACMA, Report for half-year ending, 1 July 1872. At a tea in September at the People's Gardens, Willesden, to celebrate the amalgamation, one speaker looked forward to the day 'when the association should extend beyond London', and a discussion followed of the 'differences' between the East and West End societies. See *Bee-Hive*, 14 September 1872.
47. FSOCM, *MR*, August 1873.
48. ACMA, Reports for half-years ending 5 January 1874 and 6 July 1874.
49. FSOCM, *YR*, 1876.
50. T. Scarrott to A. Yule, 5 November 1873, reprinted in FSOCM, *Special Circular*, 17 November 1873.
51. FSOCM, *YR*, 1877.
52. For example, before the FSOCM established its London branch in 1876, it complained that 'the majority of our members going to London have entered the Alliance Society', rather than pay the 14 shillings entrance fee required by the West End Society. And in Birmingham, the secretary of the FSOCM branch told Sidney Webb in the early 1890s that 'most' of that City's Alliance members would be eligible to join the FSOCM, although that had not been the case in earlier years. Webb concluded that 'It looks as if the skilled men had been attracted away from the [FSOCM] by the lower contributions of the Alliance'. See letter from A. Yule to T. Scarrott, 15 October 1873, reprinted in FSOCM *Special Circular*, 17 November 1873; WTUC, A, XXII, 1, 26–31; FSOCM, *MR*, December 1875.
53. FSOCM, *YR*, 1862, 1890; ACMA, *YR*, 1882, 1890.
54. Rule 22, clause 5 (date of rule-book not known), reprinted in FSOCM *MR*, August 1877.
55. Rule 44, ACMA, 1884 rule-book. Rule 44 also proscribed the acceptance of price or lump work in an exclusive day workshop, or of day wages below the established level of the shop.
56. FSOCM, *YR*, 1875–91; ACMA, *YR*, 1875–91.
57. WTUC, A, XXII, 1, 26–31.
58. ACMA, *YR*, 1890.
59. In these four years, the FSOCM grew from 1126 to 1374; the United Operative Cabinet and Chair-Makers of Scotland grew from 714 to 1252; and the Alliance grew from 1101 to 5380. See FSOCM, *YR*, 1887, 1891; United Operative Cabinet and Chair-Makers of Scotland, *YR*, 1887–8, 1890–1; ACMA, *YR*, 1889, 1891. See also the membership estimates of six cabinet-makers' societies by T. Jelliffe, an Alliance official, in *Royal Commission on Labour*, *op. cit.*, qs.19, 892–19, 940.

2. Bondage in the Bakehouse? The Strange Case of the Journeymen Bakers, 1840–1880

Ian McKay

Throughout nineteenth-century Europe men and boys worked in bakehouses at night, often underground and for as long as twenty hours. Although bread was undeniably the most important element in the working-class diet, the men who baked it were among the most savagely exploited industrial workers: the most necessary craftsmen were paradoxically the most abused.[1] Journeymen bakers in the United Kingdom suffered from this overwork and meagre compensation as much as other European bakers, and in Ireland, England and Scotland a determined effort was made to reform the industry in the third quarter of the nineteenth century. Notwithstanding the common features of the industry in the various parts of the country, these attempts at reform met with very different results. Attempts to end night-work and establish strong craft unions in England and Ireland failed, but the Scottish bakers succeeded both in establishing strict trade union controls over their craft and in ending the system of night-work. This Scottish success raises a number of interesting questions, most notably about the interpretation of crafts and of the internal stratification of the working class. E.J. Hobsbawm suggested in his classic discussion of the labour aristocracy that the frontier between labour aristocrats and others was often a geographical one.[2] But how did these frontiers come to exist, particularly within a craft which supplied a relatively homogeneous product? Why were some bakers able to sustain the sense of craft and others unable to do so? What were the consequences of such a division within the industry? How much did such a division reflect underlying economic facts as distinct from divergent ideologies and practices? By answering these questions within the context of this marginal craft, we may be able to establish the margins of craft: that strange structure, at once an objective fact of capitalism and a subjective system of symbols and values, which seems to encompass freedom and necessity, obedience to economic structures and the will to change those structures.

I

It is logical to begin with those features of the baking industry in the United Kingdom which were common to all its regions. The most essential point is that baking did not become a modern industry anywhere in the United Kingdom in the nineteenth century. The reasons for this were many. First, bread could be made in the home and the existence of household production diminished the incentive to industrialise. Second, bread had only recently emerged from a nexus of traditional constraints which, when coupled with consumer conservatism, rendered any substantial innovations with the basic product unprofitable. Third, the perishable nature of bread and the limitations of nineteenth-century distribution systems placed distinct handicaps on any attempts to create plant baking. Finally, the problem of distribution intensified the relative technological backwardness of the art itself, and the most revolutionary changes in ways of making and packaging bread were to come after 1914.

Baking was in many areas seen as a domestic task. Those listed as bakers in the occupational tables of the census numbered 48,809 in 1851 and 72,367 in 1871. This represented an increase in the ratio of bakers to thousand population from 2.38 in 1851 to 2.78 in 1871. The acceptance of public baking was hardly uniform across the country. In Scotland, such counties as Inverness, Argyll, Ross and Cromarty, Sutherland, Caithness and the Orkney and Shetland Islands in the north had far lower baker: population ratios than Edinburgh, Linlithgow, Lanark and Renfrew in the south.[3] The general 1851 English ratio of 2.36

bakers per thousand conceals the remarkable divergence between the southern and northern counties. To cite the most glaring contrast, the ratio for London of 4.48 bakers per thousand was far above Yorkshire's total of 0.89 bakers per thousand. In 1871 the ratio for England and Wales rose to 2.60, but this concealed the remarkable difference between London, the south-east and the South Midlands (all with over four bakers per thousand) and Yorkshire, the North and Wales (all with slightly over or under one baker per thousand).[4]

All authorities agree that there was a general relationship between public baking and the growth of cities, but this cannot be seen as an invariable rule. For reasons which social historians have been slow to explain, there are pronounced variations between the North and South of Britain in food preferences.[5] This would seem to apply in this case as well, since such major urban centres as Bradford, Halifax, Leeds and Sheffield (each with fewer than one baker per thousand in 1871) coexisted in the same country as Exeter, Bath and Bristol (each with more than three bakers per thousand).[6] It may be that areas in which mining is important, and which consequently have a disproportionately large male workforce, may be logical places for a sort of working-class traditionalism which stresses home baking. Whatever the reason, the contrasts are most pronounced. When the Medical Officer of Health for York was asked in 1862 to report on conditions in the bakehouses in his district, he replied that the 'Custom [was] general for householders to bake their own bread,' and his counterpart in Bradford made the same point: 'Very few public bakehouses in the town; custom general to bake in private houses.'[7]

That baking was still a domestic occupation is further confirmed when we consider the role of female labour in the industry. Although 11 per cent of the English and Welsh bakers, and 6.3 per cent of the Scottish bakers were female in 1851, and roughly the same percentages (10 and 7.1 per cent respectively) held in 1871, this did not represent a female presence in the workforce, but rather the management of bakehouses by women who worked alongside their husbands or by themselves. In Yorkshire — the area most resistant to public baking — the number of women in the industry exceeded the number of men. The transition to the modern bakery often began with modest efforts in slightly enlarged country kitchens, a domain where women naturally enjoyed a certain expertise.[8] Other transitional forms included the meal men of Dublin, who sold meal for bread to the poor, which they took to a public oven to bake themselves, or the English tradition of making the dough at home and bringing it to the baker.[9] The consequence of this domestic baking was that any large-scale price-fixing would create a movement back towards home baking. There were social limits to the profits of the industry, simply because an alternative to public baking was readily available. In fact, to the delight of William Cobbett, such a return to home baking was evident in the years of distress in the early nineteenth century.[10]

The second aspect of the backwardness of the industry is the late and uneven conquest of baking by capitalism. As E.P. Thompson has so memorably shown, many working people in the eighteenth century did not accept the application of political economy to the most important single part of their diet.[11] Although the Assize of Bread — the statute which since Henry III or earlier had regulated the relationship between the miller and the baker — was formally abolished in 1822 in London, and in 1836 in the provinces, its effectiveness as an instrument of price control had waned after Justices of the Peace were entrusted with it in 1710.[12] Bakers who found themselves bound by the price of wheat, but unprotected against changes in the price of flour, were driven into dependent relationships with heavily capitalised millers, or forced to undersell their fellow tradesmen.[13] In London by the end of the eighteenth century effective control of quality and price had virtually ceased.[14] In Glasgow the Assize was discontinued in January 1801, and bakers were left to 'furnish bread to the inhabitants at such prices as they can afford it,'[15] although the city's bakers continued to relate bread prices to the price of wheat as though the Assize were still in force.[16] The abandonment of the Assize was justified by arguments that large-scale capital would enter the field and produce cheaper bread. In fact, the realities of nineteenth-century distribution costs meant the repeal of the Assize created an influx not of men of capital, but men on the make; former journeymen attracted by the low capital requirements of entry to the trade. One reason why the abandonment of price control did not rationalise the industry was the tenacious conservatism of British bread consumers. Since the shift (much less pronounced in Scotland than in England) to white bread in the eighteenth century,[17] there were no basic changes in taste for the rest of the nineteenth century. Since bread's perishability and bulkiness favoured retailing from the bakery, and since journeys to the bakery were generally on foot, consumer preferences conferred an enormous advantage to the small local bakery.[18] With due regard for regional variations, we may still say that there was an inherent tendency to savage competition in the industry, and because of the exceedingly small costs of entry and consumer conservatism, the obstacles facing any large-scale capitalist were formidable.

Although the problems of perishability and distribution were fundamentally important, we should not overlook the fact that no technological breakthrough had occurred which could have transcended these constraints. The crucial problem here was controlling the speed of fermentation to enable a precise estimate of time required.[19] Because of the adamantine conservatism of the consumer, any factory-produced bread would have to be virtually identical in taste with the article produced in the neighbourhood bakery, but the most ambitious attempts to change the industry were based on product innovations. The Dauglish process (which 'aerated' the bread as it passed from a mixing machine under pressure) was highly sanitary, could be easily adapted to factory production, and very economical. The bread it produced was tasteless and won only a tiny fragment of the market.[20] Competition from small bakeries was the reason cited by the directors of the Aerated Bread Company of London, which used the process, for their failure to gain acceptance.[21] Mixing machines, of which the Stevens Patent Dough Making Machine was the most popular, did find a large market among master bakers, but did not radically change the trade. As Stevens himself pointed out,

> As the great majority of bakers are on a very small scale, employing only from one to two or three men, the introduction of my machine would not lead in those cases to fewer men being employed, or indeed in any other cases except in the very rare ones of large establishments, where ten men and upwards are employed.[22]

There were, finally, incessant efforts to reform the design of ovens and signal advances were registered in providing indirect heat and thermostatic control.[23] No technological breakthrough dislodged the local bakeries from their privileged position, and this system of small, competitive producers was self-perpetuating, since small masters, who could enter the trade cheaply and survive in it through heroic feats of self-exploitation, could always undersell the heavily capitalised bread factory.[24]

That plant baking emerged at all was the consequence of the improvements in the transportation system brought about in the twentieth century. This brought within reach a greater number of potential customers, and gradually undermined the position of the small masters.[25] Large capitalists would succeed here only when monopoly capitalism had transformed the market. No such major transformation occurred during the nineteenth century, and the relative backwardness of the baking industry had been established beyond doubt. In this context, the way of making bread changed little in hundreds of years. In England, the making of bread began between eleven and twelve o'clock in the day, when the 'ferment' was made. This ferment was made from potatoes, water and brewer's yeast, and was left for six or seven hours. It was ready for use once fermentation ceased. Bakers 'set the sponge' by mixing the ferment with one-quarter of the flour, a demanding task which required two men for two sacks. The 'sponge' so made was left until midnight, during which time it rose twice; then the remaining three-quarters of the flour was added, together with much water. Now the hardest labour began, that of making the dough (only outside the industry was this called 'kneading'). After the dough was made it was left to stand, from half an hour to two hours, and then it was weighed, moulded and put into the oven, where it stayed until the baker's judgement and local tradition decreed that it was well baked.[26] This way of making bread led to a rather long day, and since bakeries made between two and four batches a day, journeymen bakers found themselves hard driven. In some bakeries, both foremen and journeymen spent the day-time making deliveries.[27] Some journeymen put in an eighteen-hour day, and casual workers hired to meet the weekend rush might work straight through from Thursday evening to Saturday afternoon.[28]

It is not difficult to understand why working conditions in the baking industry were so wretched. There were 'natural' reasons for night-work, such as consumer preference for hot bread and rolls in the morning, and equally obvious reasons why labour in baking bread tended to be long and arduous. It is hardly surprising, given the chronic overcrowding of the trade, that profit margins were slashed to a minimum by competition, and that consequently master bakers were unable to pay decent wages to their men. And these considerations, which all follow logically from the underde-veloped, *petit bourgeois* nature of the industry, make it easy to understand why trade unionism in the industry as a whole was immensely difficult to build and sustain. Workers who were able to overcome the formidable obstacles placed in their way by the divisiveness of small-scale production and the very important fact that many of them would become small masters themselves,[29] found themselves confronted by employers who could not organise themselves into coherent bodies. Demands for the reform of the industry could scarcely succeed when those who controlled it could not even defend themselves against their own price-cutting and sharp practice.

These common factors suggest that the baking industry, where it was accepted as part of social life, would have created much the same sort of trade unionism in all parts of the country. This view is mistaken, and to see why we must examine the different industrial structures which emerged in Scotland and England, and specifically in Glasgow and London.

II

Within the baking industry, London stood as the most extreme example of the collapse of price and quality control and the underselling economy this initiated. The London bakers, Sylvia Thrupp has shown, were divided between white and brown bread bakers, and in the eighteenth century by a widening gap between those prosperous bakers who had invested in milling, and poorer bakers who tended to become more and more indebted to flour merchants.[30] As early as the seventeenth century, poorer bakers were losing their autonomy by falling into the hands of millers, men of capital who had first made their money in the meal trade.[31] It is not clear when the competition in the trade assumed the extreme form that would be well known by the 1830s, but there is no question that the abolition of price-fixing 'wrought a revolution which transformed baking into one of the most depressed, overcrowded and unremunerative trades of the day.'[32] Underselling probably began as early as the 1780s, but shortly after the repeal of the Assize the industry was swamped. 'Some years ago,' one London baker recalled, 'the greater portion of the London trade was in the hands of a comparative few. Large trades (rarely to be met with now) were done, and with, as I think, the advantages and protection afforded by the Assize, large profits were won.'[33] Once opened to full competition, baking initially was an attractive field. Where the full-price bakers under the Assize had granted large amounts of credit, the new 'undersellers' insisted on cash sales. While the full-priced trade had delivered to the door, the undersellers sold bread in the shop.[34] Undersellers also took advantage of shrewd investments on the flour market. They took over the trade, and the full-price bakers retreated to their sanctuary in the West End. At a certain point, however, the trade became glutted; by the 1840s, one baker recalled, one saw an 'insanity of competition, unexampled in any other trade.'[35] New 'half-penny-under men' or 'cutting' bakers now undersold the undersellers, bread was hawked on the street by unemployed journeymen, and stale and cheap bread was sold at chandlers' shops. By 1862 one-sixth of London's

estimated 2500 bakers were said to belong to the full-price
trade, and one-sixth to the 'cutting trade', with the remainder belonging to the legitimate undersellers. This became a
nightmare for many small masters, who were constantly
pressed to reduce their prices; one 'cutting' baker in the
neighbourhood could force all the bakers down to ruinously
low prices.

The London bread economy was governed by an implac-
able logic. By the 1850s and 1860s the legitimate undersellers were struggling, through their weak and ineffectual
protective associations, to regulate bread prices, but to no
avail. There were grave difficulties in setting a standard
price when there were three classes (the aristocracy and
wealthy, the retailing and better artisan class, and the
labourers and poor) each needing different qualities of
bread, one master baker noted.[36] Had these been separate
and distinct markets, each served by an easily recognised
and distinctive product, the master bakers might have been
able to set up uniform price-lists. In practice this was
impossible. Moreover, since many London bakers were tied
to milling concerns, bakers who insisted on cash only in their
stores were able to secure long lines of credit with millers,
who served in effect as the bankers of the trade. But the
close association with millers meant that the price of bread
had to reflect the violently fluctuating price of flour. There
were many master bakers who speculated from necessity and
lost everything, and others defrauded the milling companies
by defaulting on payments. But before such frauds were
exposed they had the effect of depressing all the prices
within the neighbourhood. J.C. Dwarber, one of the most
perceptive small masters, argued that the many evils of the
London trade could be cured if the millers gave credit only
to those bakers who were demonstrably capable of
repayment.[37]

It was clear that many of London's bakers survived
only by adulterating their products with alum, chalk or
Derbyshire stone.[38] Flour factors were thought to have
dotted the city with 'half-penny-under' shops by paying for
their flour and providing such shops with an adulterated
product. Such adulteration made it impossible to establish a

system of grading, since loaves of bread similar in appearance and taste might be quite dissimilar in their nutritional
value. Working-class consumers, driven to chandler's shops
where the worst bread was sold because such shops granted
credit, bought bread of little nutritional value. It is a history
which must give the most determined 'optimists', who have
made such capital out of the London bread market, whose
prices they have illegitimately applied to other regions,
grave cause for concern.

Faced with the savage competition of the industry,
London masters could either overwork their journeymen,
adulterate or tamper with their bread, or declare bank-
ruptcy. They had virtually no room in which to manoeuvre.
One example may be given which shows the precariousness
of the small London master: in the typical shop it was the
universal custom for the man who stirred the 'sponge' to
wash his arms in a pail of water and leave the water standing
until the dough was made; this water was then thrown in the
dough. Saving the water was a trifling detail, but it meant
saving a halfpenny a week, and no small master could afford
to overlook that.[39] One London master baker revealed the
slim profit margins of the London industry in a paper read
before the Master Bakers' Trade Protection Society in 1870
which estimated the average London bakery doing twenty
sacks a week could clear £3 50s, from which the baker had to
pay his own living expenses and keep his shop in good
repair. This was a tight margin. If flour was priced at 38
shillings bread could not go underneath 6½d — and yet
there were undersellers who pushed it to 4½d. We understand why the incessant struggle to survive led one prominent writer and master baker to look forward to the day when
large-scale industry would do away with the type of small
bakery he himself operated.[41]

In this context the savage exploitation of London journeymen was hardly surprising. High property values in the city
drove many of the bakehouses underground, and the
desparate plight of the masters precluded much attention to
ventilation or cleanliness. The long hours and night-work to
which the trade was everywhere prone were exaggerated by
the competitiveness of the London trade. Journeymen

generally slept in the bakehouses, even if this meant using the bread board for a bed and a baking pan or a flour sack for a pillow.[42] In the bakery which made oatmeal for the Royal Family, five of the nine hands slept on three bedsteads which rested on top of the oven. There was no floor between the top of the oven and the bedstead, and one of the bakers understandably commented that 'The heat of the oven on which these beds stood was very oppressive.'[43] Dust from the flour and fumes from the oven caused severe respiratory problems, and the long hours of labour drove many journeymen to an early death. But what particularly horrified the mid-Victorians (and, less forgivably, the Fabians) was the unsanitary nature of underground bakehouses.[44] Many London bakehouses were filthy, smoky coal-holes, with cobwebs hanging from ceilings and seepage from water closets on the floor. Tremenheere uncovered one case of a master baker who used the same oven to bake bread and dry dyed hair for wigs.[45] The details of the trade in the 1862 Report of Tremenheere, said Marx with black humour, 'roused not the heart of the public but its stomach.'[46]

Wage rates varied with the type of small master, since all collective control over the London trade had collapsed. The 1862 Report presents an optimistic portrait of such rates, since it is systematically biased in favour of the better established and more reputable shops. John Lilwall, in *Bondage in the Bakehouse*, claimed that the wages in the industry ranged from 12 to 18 shillings per week; master bakers, in reply, responded with estimates from 14 to 32 shillings.[47] This debate, which is treated in an appendix, is unlikely to be fully resolved, since by definition the unregulated economy of London baking was without a standard for wages or hours. It is certain that many of the poorer masters did away with first, second and third hands altogether, and paid miserable wages to children and foreigners.[48]

This was an industrial structure, and had one been able to alter one of its principal elements one would have changed them all. The enduring low level of technology in baking encouraged an influx of undercapitalised journeymen as

small masters, who in turn guaranteed by their competition
the failure of bread manufactories. The proliferation of
small masters created the need for adulteration.[49] Barbaric
working conditions made it possible for small masters to eke
out a living, and the loss of control over the trade made it
impossible to change these conditions through negotiation.
Once the decision was made to place the vital staple of
popular diet in the hands of the economy of competition — a
decision which was made by the state — all else necessarily
followed. It is difficult to imagine a more complete contrast
to the 'moral economy' of eighteenth-century England than
this Hobbesian world of small masters, at once proliferous
and deprived, united only in their desperate efforts to avoid
financial ruin.

London was the extreme case of the collapse of all
controls over the baking of bread, and its brutal economy
stands as a paradigm of the social consequences of capitalism. The history of baking in Glasgow is altogether
different. Although effective control had disappeared in
London by the end of the eighteenth century (and well
before repeal of the Assize), for years after the Glasgow
bakers conducted themselves as though the Assize were in
force. The price of Glasgow's quatern loaf was regulated by
the city's Incorporation of Bakers until at least 1834.[50] The
Incorporation took most seriously the legal right to exclusive
trading privileges, and in 1801 prevented an 'outentowner'
(outsider) from establishing a business manufacturing
bread.[51] The dependence of early London bakers on
large-scale capitalist millers noted by Thrupp can hardly
have been duplicated in Glasgow, since the Incorporation of
Bakers owned the largest and finest flour mill in Scotland,
which it still retained as late as 1884. [52] The repeal of
exclusive trading privileges was swiftly followed by the
establishment of relatively large baking concerns. Throughout the middle decades of the nineteenth century Glasgow
and Edinburgh master bakers were able to set collectively a
price for bread.[53]

It is of utmost importance to remember that the underselling economy never took hold in Glasgow. This, however,
would not provide a complete explanation for the much

larger Glasgow bakeries. The reason for the peculiar path followed by Glasgow (which has characterised its baking industry well into the twentieth century[54]) is its distinctive pattern of urban growth. Glasgow was the most over-crowded major city in Britain. Although the population had increased fivefold between 1801 and 1861, the whole of the Glasgow of 1775 remained geographically intact.[55] Glasgow's density of population has long been noted by students of urban development.[56] Using the English statutory defini-tion of overcrowding (more than two persons per room), Eric E. Lampard notes that in 1901 London had 16.0 per cent overcrowding. Glasgow by contrast had 55.1 per cent, and Dundee and Kilmarnock were even higher.[57] The Scottish preference for barrack-like stone tenements, many containing 'single-ends', was no doubt partly responsible for these peculiarities.[58] The consequence of this was that bakeries, without any significant improvement in their methods of production or distribution, could economically produce for a far larger number of people.[59] There were also distinctive patterns of bread consumption. Throughout Scotland and the North oatmeal must be considered as one principal alternative to bread as a staple of popular consumption, and possibly as a consequence there was not the same insistence in Glasgow on white bread characteristic of demand in London.[60] Since the use of alum as an adulterant was principally aimed at producing an illusion of whiteness (and hence allowed the baker to charge more for bread made from coarse flour), it seems plausible to infer that the practice of adulteration was not as common in Glasgow, Scottish taste in bread and radically different oven design (with walls a foot thick instead of 2½ inches) meant that bread could be baked more rapidly.[61]

The absence of underselling (engendered by the distinc-tive development of Scottish capitalism),[62] and the much greater density of Scottish cities allowed the emergence not of baking *factories* based on steam technology, but *manufac-tories* based on the congregation under one roof of a larger number of craftsmen and on a more widespread distribution of bread. The largest Glasgow bakery, Crossmyloof, was founded in 1847; by 1859 its 80 employees were baking 140 sacks of flour daily, in 38 ovens.[63] The suriving plans of the bakery demonstrate the size occupied by the stables, and hence the extensive distribution undertaken.[64] This large Glasgow bakery, hailed as a 'model workshop' by Charles Lang, secretary of the Scottish bakers' union, was bright and well-ventilated.[65] Several of the 156 other bakeries in Glasgow in 1857 were almost as large, although of course the neighbourhood bakery was still present.[66] The critical point is that the presence of the larger baking enterprises gave the Glasgow trade a stability and cohesiveness which made collective bargaining possible.

Compared with the miserable working conditions of the London bakers, Glasgow journeymen had reason to be thankful. Through determined public campaigns and trade union struggles, the Scottish bakers were able to end night-work in 1846, which excepting minor lapses was an achievement which was successfully defended until 1877. In contrast with their London counterparts, Glasgow bakers abolished the living-in system and secured a rate of remuneration in excess of that which obtained in London (from 18 to 22 shillings a week for the twelve-hour day, compared with the same sum in the *best* bakeries in London for a sixteen- to eighteen-hour day),[67] which is all the more remarkable when we consider that the general tendency of Scottish wages was below those in England.[68]

The elements of the exploitive and self-perpetuating system of overwork and underselling in London were absent in Glasgow. The presence of well-capitalised bakeries diminished the problem of overcrowding (although it did not, as we shall see, eliminate it). The absence of endemic underselling removed the need for widespread adulteration, and since the production of bread never became the province of the disorderly competition of London it was less necessary to rely on overwork. This distinction between the two urban economies illustrates the differences which may exist even within the same trade and within the same basic realities of capitalism. But these local differences, which created such different milieux for working-class struggle, were not so wide that bakers pursued merely local objectives or felt themselves to be part of merely local crafts. Quite the contrary; Scottish journeymen, the élite within the industry, sought to reform it throughout the country.

III

When we look at the common patterns of trade unionism in England, Ireland and Scotland, it is useful to remember how interlocked some of these groups of craftsmen were. Many Scottish bakers emigrated to London and (with a speed which some other workers found distressing) became foremen in the best bakehouses.[69] They were labour aristocrats in the sense that they identified themselves with the political movements associated with the upper stratum of the working class. They expressed a sort of moral superiority when they implied that their own success in ending night-work was the consequence of dedication and temperance. We shall note that the success of the Scottish bakers was contingent on the export of ambitious Scottish journeymen to London, and that in this indirect sense the Scots were able to defend their gains only at the expense of other workers. None the less, the Scottish bakers were outward looking, even evangelical, in their concern for bakers in other areas. Without them it is difficult to imagine any of the public campaigns against night-work or the passage of the Bakehouse Regulation Act in 1863.

In a certain sense, the course of trade unionism in England, Ireland and Scotland during the third quarter of the nineteenth century was uniform. In all three countries an attempt was made by bakers to end excessive hours in the 1830s: in Ireland by agitation for the Sunday Baking Act of 1838 and large public campaigns in Derry and Belfast which succeeded in eliminating night-work;[70] in England by a large, unsuccessful strike in London in 1836.[71] Early Scottish agitation was centred on Edinburgh and succeeded in ending the system of boarding and lodging in the houses of employers, which the bakers denounced in an eloquent memorial as 'unjust in principle, pernicious in practice, and unscriptural in its operation.'[72] Even in the 1830s the Scottish bakers enjoyed greater success than those of other cities. More sustained attempts were made in the 1840s to establish permanent union structures. In Ireland a revived campaign against night-work swept the country in 1842, but collapsed equally swiftly. An extemely impressive campaign, which enjoyed the support of some Chartists, sought to end night-work in England in 1848, but the attempt to have Parliament legislate on the question offended John Bright's free-trade sensibilities and reminded other members of the national workshops of Louis Blanc.[73] The most important and dramatic victory was that of the Scottish bakers, who in 1846 formed the Operative Bakers' National Association of Scotland, and through a strike and a large public movement secured the end of night-work and the ten-hour day.[74] This event, celebrated at countless soirées in the 1860s, was subsequently an important aspect of the union's mythology.[75] This was indeed an event to remember, since it is clear that with the exception of one or two localities, night-work was eliminated in Scotland. However, many other goals of the union were lost in an unsuccessful strike in 1857. This strike sought to control the number of apprentices and to establish a common wage rate of 22 shillings; it was a bitterly fought affair. Glasgow employers set up an alternative house of call for country blacklegs, and journeymen bakers in response established a cooperative baking company with the full support of the Trades Council.[76] Although the union continued to insist that no member should work beside a non-member and that none should accept a situation, 'the character of which is unknown or doubtful, without first inquiring of the officers of the nearest branch,' it appears that the society's right to impose a closed shop was lost through legal action as a consequence of the strike.[77] The long-term consequences of this defeat, and particularly of the failure to restrict the number of apprentices, were to be most serious.

Perhaps as a consequence of this reversal, Scottish bakers took the initiative in attempting to secure the reform of the baking trade by Parliament, and this campaign dominated working-class agitation in the trade from 1859 to 1864. Under the leadership of John Bennett, former secretary of the Scottish union, bakers throughout the country united to demand an end to night-work. It was during this extraordinary campaign, immortalised by Marx in *Capital*, that the reform of the bakehouse became a popular issue for members of the middle class who were concerned with sanitary reform, temperance, and adulteration.[78] The Bakehouse Regulation Act of 1863, which prohibited youths under 18 from working in bakehouses from 9 p.m. to 5 a.m. and made inadequate provision for bakehouse inspection, made a most important change in the trade, particularly in areas where unions could report delinquent employers to the the health authorities.[79] Although much of the testimony which was collected by Tremenheere and others related strictly to London, in fact the Act must be seen as a Scottish measure, instigated by the London Operative Bakers' Vigilance Committee, but inspired and led by Scots. This legislative campaign had an impact on trade unionism, and the Amalgamated Union of Operative Bakers was founded in Manchester partly as a result of this agitation, in 1861.

A final wave of trade union activity swept all three countries in the early 1870s. The English union became uncharacteristically active with its locals fighting successful strikes for shorter hours in Bath, Bristol, Leicester, Nottingham, Liverpool, Manchester and district, Derby and the Potteries.[81] In London, where four unions (the AUOB, a small local union, the London Union of Operative Bakers and a small German society) reflected the divisiveness of the industry, a strike in 1872 failed dismally to achieve any alteration in nightwork or wages.[82] There was a great deal of activity in Scotland, and a noteworthy re-organisation of the union, but growing dissension on questions of an aliment scheme and local control paved the way for the collapse of the Scottish craft in 1877, when employers reinstated night-work and the union collapsed. In Ireland a remarkable movement of bakers in 1872 sought to build the Irish Bakers' National Union. With 26 of the 32 counties within the fold, the union seemed a decisive advance over the small local societies which had successfully maintained high craft standards in Dublin. It was defeated, however, in a bitter eighteen-month strike in Dublin, during which the employers set up their own house of call and recruited Scottish and English blacklegs. The union collapsed in 1876.[83]

This bare summary of events scarcely indicates the byzantine complexity of the evolution of these local bodies, and naturally fails to describe the passions and hopes which such trade unionism aroused among these workers.[84] It is sufficient, however, to demonstrate several key patterns. The most obvious is the failure of all unions outside Scotland to become effective regional bodies. The English AUOB, although it just managed to survive the 1870s, was hardly anything more than a clearing-house for information about local societies and a modest benefit society. Its organisational activities were confined to the Manchester area. Having implicitly abandoned the struggle against night-work, the AUOB was a union in name only, a New Model which possessed the form but not the substance.[85] The London Union, a quaint little organisation, may be kindly said to have had a faint imprint on the London trade, and the German society, which no doubt comforted many of the immigrant bakers in the metropolis, even less. The Irish union, like the earlier campaigns of 1842 and 1859–60, blazed across the firmament but left little trace behind it. Only in Scotland did bakers possess the trappings and the reality of regional craft unionism.[86]

This history of the bakers' struggles follows the main lines of working-class history in this period. This is clearly shown by the very deep response touched off among the bakers, for understandable reasons, by the nine-hours' movement. Many were the speeches made by journeymen bakers in 1872 chiding fellow craftsmen for not joining the general movement for shorter hours, and 1872 marked the high point of militancy in the craft.[87] The defeats of the later 1870s were clearly related to the flooding of the labour market in the recession. So grave was this problem in Ireland that the Bridge Street Society of Dublin, an ancient and conservative society, paid its members £15 if they would guarantee to go away for five years.[88]

It is not difficult to understand the very limited success of trade unionism in this industry. The journeymen bakers faced a highly ambivalent situation. They did not defend an exclusive skill. Although many tried to argue that the craft secrets of the baker were many and deep, or that bakers required a profound knowledge of chemistry and physics for their craft,[89] such claims were obviously overdrawn. Many housekeepers had the 'arts and mysteries' of baking well within their grasp. Moreover, apart from the foreman and the first or Scotch hand, who supervised the work and made the decisions, the labour of the baker was not skilled in any ordinary sense. Bakers worked in filthy environments and for employers who were often dishonourable in their dealings with the public. They were, in short, the sweated labourers or 'white slaves' immortalised in the countless tracts of the bakehouse reformers. On the other hand, bakers worked in an ancient trade, could harbour reasonable expectations of setting up on their own account, were remunerated above the level of the labouring poor, and were tied to a wider context of respectability through journeymen's clubs, benefit societies and recreational activities.[90] Given the immense difficulties of organising such a trade, one is forced to remark not on the failures of the journeymen bakers but on their persistence.

This split between the two interpretations of the bakers' situation is not unlike the division in other trades between honourable and degraded sections. In the baking industry this took the unusual form of a marked contrast between Scotland and London, each representing alternative possibilities. Of the experience of the journeymen bakers in London, enough has been said to show that the anarchy of production eliminated all possibility of an effective trade union. The history of the bakers' struggles in London was one of unmitigated failure. But it remains necessary to look more closely at the other pole, in Scotland, not only to see how the objective preconditions of craft control were in fact acted upon and incorporated in practice, but also to attempt to explain why such an impressive craft union, seemingly so much stronger than those elsewhere, suffered such a total defeat and disorganisation in the 1870s.

IV

The emergence and early success of the Scottish union was the consequence of the structure of the baking industry, particularly in Glasgow. The absence of underselling, the existence of high urban densities suitable for bread manu-fatories, and certain aspects of ways of making Scottish bread have all been seen as important parts of an explana-tion of the genesis of this distinctive structure. The larger scale of Scottish baking enterprises was far more congenial to successful craft unionism than the classically 'artisanal' bakeries of London. Yet these structural facts must not be turned into a fundamental or prior explanation to which all else may be reduced. It is important to remember that the Scottish bakers were not merely economic men. The petition of Edinburgh bakers in 1835 against the board-in system defended a concept of man as ' "the noblest work of God", capable of spanning immensity and taking knowledge of all the innumerable worlds that adorn the ethereal dome, with powers of mind qualified to "measure an atom, and to gird a world," ' and indicted the masters for unscriptural conduct.[91] Their outlook was one of a religious and passionate rationalism. A.J. Hunter, one of the union's most important leaders, argued that it was the 'want of moral culture' that hade made it possible for employers to look upon bakers as 'mere *chattels, things* belonging to our employer, to be used by him as he thought *fit*, to advance his interests, without once consulting our opinion in the matter'.[92]

If we do not consider this question of what the Scottish bakers believed, it becomes difficult to understand why they so often served as missionaries in England and Ireland. The three most important leaders of the union, A.J. Hunter, Charles Lang and John Bennett, all wrote important letters to other unionists urging them to stand firm. Hunter even initiated a correspondence with W.N. Hancock of the Dublin Statistical Society after reading about Hancock's role in the campaign against night-work.[93] Lang urged the English bakers that 'If they would only approach their problems firmly, they would much easier overcome them,'

and typically told London bakers to hold out for a set
starting-time of 5 a.m. rather than a proposed system of any
twelve hours.[94] The most conspicuous example, of course, is
John Bennett, who single-handedly coordinated the London
bakers' struggle against night-work in 1859 and 1860 and
remained to take a leading role in the troubled affairs of the
AUOB in that city. Bennett sought to replace the London
pubs — the traditional houses of call for bakers — with
registry offices, and in the one he did establish, itinerant
bakers could browse through a library stocked with 100
volumes from the Society for Promoting Christian
Knowledge.[95] Such men played a crucial role in the affairs of
the Glasgow Trades Council and were found among the
most politicised of its members.[96]

To grant an importance to these concepts of independence
and self-culture which the Scots so arduously defended,
must not lead us to the mistake of considering such values as
part of an autonomous 'culture' which existed apart from the
realities of material life.[97] These symbols and values were
widely diffused and were not the exclusive property of the
labour aristocracy.[98] But it is even more important to recall
that the differential access to such 'symbolic capital' which
the Scottish bakers enjoyed was the result of objective
economic structures, inherited from the past and created by
the uneven development of capitalism. That the Scots
invented themselves as craftsmen — with their own legendary tales of deliverance, and their special language of
resistance — is clear; that this prompted them to reconstruct
the industry, with partial success, is beyond question. But
what is equally apparant if that this will to change the
structures of baking was itself sharply constrained and
structured by the limits of the possible, in this case so sharply
defined by the different kinds of economic structures the
bakers sought to change.

We see, then, that 'craft' is a kind of strategy, comprising
two steps: the first of these is the translation of real or
supposed skills or qualities into labour scarcities. The second
is the defence of these labour scarcities and the widely
diffused values to which they give preferential access. This
relation is sequential and asymmetrical; the craftsmen only

limit or modify the structures within which they work. The Scottish bakers pursued a strategy of threatening masters with a general influx of journeymen into the trade, and of actively encouraging the growth of larger premises.[99] They thought that the abolition of night-work had promoted the growth of manufactories.[100] The Glasgow bakers explicitly ruled that exceptions to the rules on hours would be allowed in cases of masters expanding their oven accommodation.[101] Such evidence does not prove that the impact claimed by the Scottish bakers was in fact made. But it is not, *prima facie*, unlikely that masters, faced with curtailment of their production, would invest in more ovens as a consequence of a limit to hours. Nor is it difficult, recalling the interconnection of adulteration, overwork, competition and low pay in London, to concede the point that working-class activity, which eliminated two of these aspects of the structure, must be seen as at least a part of the explanation of why a stabilised bread industry persisted, although not perhaps of why it came into existence.

The presence of trade unions, many small London masters thought, was itself a stabilising force in an industry which tended toward high degrees of competition.[102] It is, of course, impossible to specify what might have changed had the trade union not existed. What we may determine is the difference organisation made to labour, working conditions and hiring practices, and the probable consequences of this for employers' strategies. The Rules for the Constitution of the National Association, in contrast with the rules of virtually all other baking trade unions,[103] established a highly disciplined and centralised institution. The delegate meeting, which exercised supreme power, met every three years; the actual running of the union was in the hands of a central committee, which was in effect run by salaried officials. Each branch (of which there were nineteen in 1860) was expected to establish a house of call in districts with over twenty members, and any local laws made by branches were to be sanctioned by the Central Board. Quarterly statements from district secretaries outlining the state of organisation were to be sent to the Central Board on pain of fines. Although in 1858 monies collected from fines were to be locally controlled, the Motherwell Convention of the union in 1870 ruled that fines should be paid into a central fund.[104]

The rules which governed members were strict. Members were to pay entry money of from 10 shillings to £2 (although how this was established is not clear), and a weekly assessment of 9d. Idle members who were jobbing one day paid 1d. Journeymen who had served a five-year apprenticeship under a regular indenture might become members within four weeks of the expiry of their apprenticeship on payment of 5 shillings to £1. Members in arrears were fined, and after ten weeks forfeited all rights to benefit. The rules governing work were particularly strict. No member could accept a doubtful situation without consulting the officers of the nearest branch, nor engage upon any other terms than by the week and for a cash wage. Payment was to be made for all Fast Days in lieu of Sabbath spongings, and any member working on a Fast Day was to receive one day's pay over and above his regular wage. No member was to work more than twelve hours per day, including meal hours, and these hours were to be from 5 a.m. to 5 p.m., with the exception (as we have already noted) of employers who could demonstrate that they were expanding their capital investment in ovens. No work under any circumstances was to be done after 10 p.m. Members moving from one district to another had to present a certificate of membership (and a statement of account) addressed to the branch secretary, and persons without such certificates were to be fined and treated as non-members. No member was to work beside a non-member.[105]

In the general context of craft unions, such working rules are quite strict (particularly those which concern jobbing), but considering the marginal character of baking, they are remarkable. They constituted no less than an attempt to create a craft through an act of collective will. Since the Scottish union was able to organise about four-fifths of the journeymen bakers in its hey-day, it exercised a virtually complete control over the working rules of the trade. These rules were not lightly regarded as guidelines; the leaders of the union talked of them as laws, governing the craft in the same way that laws governed the state.[106] The records of the

Edinburgh and Glasgow branches suggest the workings of a police court more than the customary procedures of unionism. In Edinburgh members were regularly summonsed before the branch meeting for violating the working rules, and many were expelled.[107] In Glasgow a paid local committee issued summonses (for which any member bringing a charge had to pay 2 shillings), heard cases, and levied fines. The committee actively encouraged members to bring cases against one another by bringing in a policy in 1870 of giving half the fine taken from the guilty party to the informant. (If the case was lost, however, the informant had to pay the cost of the summons.) The Glasgow branch tried hundreds of cases a year. Among the most common charges brought against its members were working with non-unionists, working past the agreed time, failing to report infractions, rowdy behaviour and failing to job when having contracted to do so. Such a court, which was structured by the craft, equally sought to structure the craft in turn. This little world of laws and summonses, fines and expulsions, pleas and verdicts, governing to the last detail the rules of work, is the molecular structure of that image of self-culture and independence which the Scottish bakers so proudly showed to the world.

How do we explain the demise of this vigorous craft? The accepted explanation is the displacement of skilled workers through machinery.[108] This interpretation does not make sense as it stands, however, since no machines had yet been invented which would produce normal bread and thereby displace manual workers. The explanation of this sudden and dramatic reversal must take account of more factors.

The fundamental change which helps account for the collapse is the revival of competition. It cannot be proved, but a plausible case may certainly be made that the tight regulation established by the bakers on the labour market discouraged competition by standardising labour costs and thereby preventing underselling masters from using over-work as a means of economising in an essentially labour-intensive industry. However, the journeymen bakers could have no guarantee that the competitive economy could indefinitely be so discouraged. The weakening of the standard rate for wages and the limited introduction of machinery meant that the large bakeries now sought to undermine competitors by cheapening labour and lengthening hours. One interesting survey of the collapse by Robert Wells emphasised the role played by competition for the best distribution networks.

> The condition of the journeyman baker in Glasgow is somewhat similar to that of his London brother; and it is surprising to see how they have fallen into such a low, depressed state, when we consider the high condition, they held as a body of tradesmen only twelve or fifteen years ago. At that time, Glasgow was looked up to as a model of excellence — so much so, that all the provincial towns in Scotland used to vie with each other in trying to imitate her. Now a cruel magician's wand has been waved over the place, and all is changed: the smallest town in Scotland can show a cleaner sheet. . . . And this has come about, I think, through the amount of competition for work that exists to drag men into the mire. The factory system is carried on in Glasgow to a far greater extent than in any other town in Great Britain; and as most factories distribute their bread in vans through the different districts, those men who can get their bread into the market first, have the best charce [sic] of disposing of it. One factory started to get their bread out an hour before another; here is where the thin edge of the wedge was inserted. I can assure my readers that the whole of the wedge is now in with a vengeance, and employers and middlemen would not hesitate to drive another wedge in if it could be managed.[109]

It was within this context of revived competition that the dough-making machinery assumed importance. Charles Lang noted that the machinery would not be opposed by the journeymen bakers *per se*, but its very limited efficiency necessitated an extension of hours within bakeries using it because only this way could it compete with hand labour.[110] Thus the collective unity of the employers, an inheritance from the corporatist institutions of the burgh and dependent on a rough equivalence of labour costs, was destabilised by machinery not because as such it made the manual work obsolescent, but precisely because it did not.

This provides a long-term explanation for the collapse of the union, but it leaves the problem of why the bakers were unable to conduct a more successful defence of the privileges which they had preserved since the 1840s. The context of the labour market in the 1870s must be borne in mind, since after 1874 certain sectors responded to the recession. Unemployed and casual workers were naturally attracted to so simple a craft. But important as this is, it does not explain why signs of chronic overcrowding were noticeable as early as 1870. In that year Lang described the problem in one of his noteworthy quarterly reports:

Trade continues dull throughout the Branches, so that there is a large surplus of unemployed labour; and it is very much to be regretted that, with the exception of one or two Branches, nothing has been done in the way of alleviating the hardship of involuntary idleness. Members who are supremely selfish may consider that as long as they are prosperous there is no need of improvement, but the fact cannot be ignored that whatever tends to oppress or degrade those who are suffering from want of employment must ultimately rebound against the interests of those obtaining full work. The unfortunate, when treated with neglect, are apt to reason thus: — What does it matter who are in receipt of fair wages, meal hours, exemption from night-work, &c., whilst we starve — thus opening up the channel of free labour competition for the barest necessities of life, in exchange for unlimited hours and the attendant train of consequences.[111]

This analysis of the structure of the labour market was not questioned, but there were radically opposing strategies. While the crisis of the craft was caused by the emergence of competition and the worsening of the labour market, the inability of the bakers to meet this challenge must be related to their failure, for understandable reasons, to create an effective system of apprenticeship.

The peculiar position of Scottish bakers *vis-à-vis* England, and particularly London, has already been noted. Scottish journeymen flocked to London, a fact which opponents of reform liked to throw in the faces of trade unionists. When John Bright urged Parliament not to interfere in the affairs of adult male labour by legislating an end to night-work, he noted that 'the clients of the noble Lord [Lord Grosvenor] were grown up men, and they were not ordinary men, for they were Scotchmen.' *Reynolds' Weekly* responded to this by noting,

Many of the London bakers are Scotchmen. We grant the statement true; and why do the Scotch bakers come to London? Because in Scotland there is no employment for them; the master bakers generally employ apprentices, and apprentices only — an additional reason why the 'grown up men' in London should have their position and circumstances considered.'[112]

There were immense obstacles to an effective apprenticeship system in a semi-skilled trade. Although there are sources which indicate an early attempt by the Scottish unionists to regulate the number of apprentices,[113] in 1860 it was reported that '[T]he usual number is two apprentices to three journeymen, but this rule is not strictly kept.'[114] Scotland served as the training ground of the best-trained London journeymen, since Scottish apprentices coming to London stood an excellent chance of becoming foremen and ultimately employers. This fact explains why the proportionate number males 25 years and under in Lanarkshire was roughly twice the size of that in London, in both 1851 and 1871.[115]

This peculiar labour market explains why the Scots were able to create labour scarcities. Had all the apprentices stayed in Scotland the number of journeymen would have doubled every three to five years with the entry of boys out of their time. Although the Scottish union did have strict apprenticeship regulations, it placed no restrictions on numbers, seeking instead to have them regularly bound and insisting on this for admission of men into their society. The origins of this strategy remain unclear, but it may well be that the traditions involving apprentices were of such longevity in the trade that the union did not question them. The association had already confronted the problem of a surplus of hands in 1853, but it formulated a policy not of apprenticeship restriction but of assisted emigration. Lang himself favoured a strategy of an aliment scheme for out-of-work members, but this foundered on the unpopularity of increased dues. The obvious underlying reason for this reluctance to enforce a restriction of numbers is that the skills of baking were too easily acquired to make such a restriction enforceable. The association was spared the full effects of unlimited apprentices through the successful operation of the system of out-migration, but eventually the recession destroyed this structure.[116] When the Glasgow

branch appointed a special committee to examine apprenticeship in Glasgow, it learned that of 253 apprentices in the
bakeries organised by the union, only 65 were bound to 188
unbound. The unbound apprentices — 'halflins' — indicated
just how weak the union had become in regulating this
essential aspect of the craft.[117] After the collapse of the
union, Carolyn Martin estimated that in Edinburgh there
were two or three apprentices to every journeyman.[118]

The imposing structure of craft controls created by the
Scottish bakers was less strong than it appeared, because it
was predicated on the continuing ability of London to
absorb a steady influx of Scottish journeymen. The ability of
the Scots to create their craft was, in part at least, a product
of a passing disequilibrium between the two labour markets
which was gradually disappearing by the 1870s. It was this
internal weakness which explains why the best organised and
most articulate craftsmen in the British baking industry
should have been so quickly reduced. Even without the
recession of the 1870s the labour market would have become
overstocked, but the general crisis of unions in that decade
meant that the bakers' fragile apprenticeship regulations
would inevitably crumble.

V

Although objective economic circumstances were indispensable in assisting Scottish journeymen to create and defend
their craft, the unremitting zeal and fervour of this élite in
attempting to extend their privileges to other bakers went
far beyond a narrow concern for economic privileges. Their
activities can only be understood as attempts to reform the
structure of baking as a whole. Learning to play a game
whose rules they did not invent, the Scottish bakers did in
fact change the structure of their trade by ending the
traditions of night work and low pay. If we take a national
perspective, the Scottish achievement would appear far
more limited. Because of the particular circumstances of its
genesis, the craft which the Scots created could not easily be

exported. Contrary to expectations, the Scots did not pass from their objective privileges to an exclusive strategy, but sought to bring the bakers up to the level of other crafts. But their ability to do so was sharply limited by the dependence on the emigration of journeymen, and in this sense the Scottish bakers were far more vulnerable than they appeared. Yet in another sense the Scottish bakers did transform baking, by providing the main impetus for the passage of the Bakehouse Regulation Act in 1863 which restricted night-work by juveniles. The ambitions of this curious élite were thus partially realised.

APPENDIX: WAGES IN THE BAKING INDUSTRY

If we take Baxter's figure of 28 shillings per week, which Hobsbawm uses as one possible measure of the bottom limit of the labour aristocracy, we may establish that in the baking trade only foremen and exceptional workmen may be considered labour aristocrats in this strictly economic sense. Certainly, the percentage was far below the 11 per cent of British workers said to earn this amount. However, consistent and thorough data — showing the hours, wages and perquisites of each type of bakehouse worker — is not available. The many London estimates presented in the 1862 Commission conducted by Tremenheere do not give precise details for hours, and many are presented in the form of a range of wages. Since all trade union controls had broken down in London, even journeymen bakers found themselves unable to formulate a coherent wages policy: as one organiser complained to the *Workman's Advocate*: 'it was at first intended to have gone for a uniform advance of 3s. per man on the present rate, but it was found that there were so many different rates of wages paid in London, that they had now decided to leave the amount of the advance an open question' (23 December 1865).

Tremenheere failed to give many details on the underselling bakers, and consequently his wage data are applicable only to the respectable trade. Master bakers also argued that the perquisites attached to the craft amounted to as much as 8 shillings per week, principally in the form of free bread and potatoes, and charged that their journeymen could inflate their wages by as much as 12 shillings by illicit overcharging of consumers on delivery routes. From the seven detailed cases outlined in the 1862 Report, it appears safe to conclude that foremen earned from 26 to 28 shillings, second hands from 18 to 22 shillings and third and other hands 12 to 15 shillings in all cases leaving aside the allowances for bread and potatoes. An indication of the lowest acceptable wage for journeymen emerges from the rule of Branch no. 24 of the Amalgamated Union of Operative Bakers in London that no member of their branch be allowed to earn less than 16 shillings per week in 1871.

A wider range of wages estimates can be compiled from the labour press from 1870 to 1874, the best years for the craft unionists in the industry.

Table 2.1: Wages 1870–1874 (in shillings)

Location	Hours	Foreman	First or Scotch hand	Second hand	Third hand and others
1870 Scotland	66	24–38		Table hands: 17–26	
1871 Worcestershire	90			18 for 'tolerable workman'	
Warwickshire	90			18 urban, 16 rural	
1872 London	75–90	24–30		18–25	15–16
*London	72	30	26	24	21
Glasgow (Crossmyloof)	51	31 oven men		20 table men	
Leicester	72	30		24	21

NOTES

1. See Justin Godart, *Les Mineurs Blancs* (Paris, 1910), and A. Savoie, *Meunerie, Boulangerie, Patisserie* (Paris, 1922), for pertinent discussions, as well as the International Labour Conference, League of

Nations, Seventh Session, Final Vote, *Communications from Governments on the proposed Draft Convention provisionally Adopted by the Sixth Session of the Conference, Night-Work in Bakeries* (Geneva, 1925).

2. E.J. Hobsbawm, 'The Labour Aristocracy in Nineteenth-century Britain', in *Labouring Men: Studies in the History of Labour* (London, 1964), p.275.
3. All but five of the southern counties (Dumfrieshire, Kirkcudbright-shire, Wigtownshire, Peebleshire and Kinross-shire) had more than 3 bakers per thousand, while all but five of the northern counties (Perthshire, Forfarshire, Aberdeenshire, Elginshire and Nairnshire) had fewer than 3 bakers per thousand. Edinburghshire, with 5.61 bakers per thousand, had the highest ratio. Occupational Tables, 1871 Census.
4. Occupational Tables, 1851 and 1871 Census.
5. But see D.E. Allen, 'Regional Variations in Food Habits', in Derek J. Oddy and Derek S. Miller, (eds), *The Making of the Modern British Diet* (London, 1976), pp.135–47, and R.H. Campbell, 'Diet in Scotland. An example of Regional Variation', in T.C. Barker, J.C. McKenzie, and J. Yudkin (eds), *Our Changing Fare: 200 Years of British Food Habits* (1966).
6. The ratios for the largest provincial cities: Liverpool (2.41), Manchester (5.38), and Birmingham (2.29). With the exception of Manchester, all the English cities with ratios higher than the national ratio in 1871 were south of Birmingham.
7. *Second Report Addressed to Her Majesty's Secretary of State for the Home Department Relative to the Grievances Complained of by the Journeymen Bakers* (London, 1863), XXVIII, Cd.3091, 6.
8. Women also would find employment in the retail portion of the trade. I have found no cases of female journeymen bakers.
9. John Swift, *History of the Dublin Bakers and Others* (Dublin, n.d. [1948]), p.146; George Read, *The Baker, Including Bread and Fancy Baking with Numerous Receipts* (London, n.d.), 21.
10. William Cobbett, *Cottage Economy* (London, 1823), III, para.82: 'How wasteful then, and indeed, how shameful, for a labourer's wife to go to the baker's shop: and how negligent, how criminally careless of the welfare of his family, must the labourer be who permits so scandalous a use of the proceeds of his labour.'
11. E.P. Thompson, 'The Moral Economy of the English Crowd in the Eighteenth Century', *Past and Present* (50), 1971, 76–136.
12. Sidney and Beatrice Webb, 'The Assize of Bread', *The Economic Journal*, XIV, 54 (June 1904), 196–218.
13. C.R. Fay, 'The Miller and the Baker', *The Cambridge Historical Journal* (1923–5), 85–91.
14. W.M. Stern, 'The Bread Crisis in Britain, 1795–6', *Economica*, new ser., XXXI (May 1964), 183.
15. R. Renwick (ed.), *Extracts from the Records of the Burgh of Glasgow*, IX (Glasgow, 1914), 214.

16. Terence R. Gourvish, 'A Note on Bread Prices in London and Glasgow, 1788–1815', *Journal of Economic History*, XXX, 4 (December 1970), 855.
17. One important consumer preference was for hot bread, particularly in the morning. The poor relied on hot bread for three reasons, according to a master baker in Stepney: 'First, it supplies the place to them of hot dishes which other people can afford, they have not the means of cooking or preparing anything hot for themselves; next they say that the children don't want so much butter with it, they will waste the stale bread while they will eat all the new. Then again, it is, as they [sic] term is, "more filling," that is, it takes longer to digest, they do not feel hungry again so soon.' *Report Addressed to Her Majesty's Principal Secretary of State for the Home Department, Relative to the Grievances Complained of by the Journeymen Bakers; With Appendix of Evidence* (London, 1862), Cd.3027, 85. (Hereinafter *1862 Report.*) This demand for hot bread and rolls in the morning was one of the chief obstacles to the abolition of night-work. John Jellicoe, an Irish master baker sympathetic to the journeymen's cause, told the National Association for the Promotion of Social Science in 1861: 'As a master baker in Dublin, he would be willing to adopt day-work, but the ladies of the city should be got to give up looking for hot rolls in the morning'. *Freeman's Journal* (Dublin), 16 August 1861. The Scottish journeymen did not make very clear to outside admirers that their achievement of day-work meant the lesser availability of hot bread for consumers, and later rolls in the morning. See the remarks of Charles Lang, in *Glasgow Sentinel*, 7 April 1866.
18. Cf. Alan R. Pred, *The Spatial Dynamics of U.S. Urban–Industrial Growth, 1880–1914: Interpretive and Theoretical Essays* (Cambridge, Mass. and London, 1966), pp.205–7.
19. The introduction of compressed yeast, manufactured from strains of pure cultured yeast selected for high fermentative power, yielded the uniform results required by a scheduled programme of production on factory lines, as John Burnett notes in 'The Industrial Revolution in the Baking Trade', *Bakers' Review*, 15 June 1962, 1011. A very interesting commentary on different traditions in Scotland and England regarding fermentation may be found in John Mackay, 'On Baking Bread, With Special Reference to German Yeast', *Transactions of the Royal Scottish Society of Arts*, V, 4 (February 1860).
20. Dauglish's life and work are described in Benjamin Ward Richardson, *On The Healthy Manufacture of Bread: A Memoir on the System of Dr. Dauglish* (London, 1884). Peak Freen in Bermondsey experimented with the process in the early 1860s but was forced to discontinue because of high distribution costs. The history of 'bread reform', from such amateur crusaders as Stevens and Dauglish, to the Bread Reform League of the 1880s and the apostles of scientific baking — John Blandy, William Jago, and Robert Wells — awaits its historian.

21. 'The Aerated Bread Company', *Bakers' Record*, 5 October 1872.
22. 1862 *Report*, 47. Ebenezer Stevens, who wrote one of the most eloquent tracts against conditions in the bakehouses, was himself the subject of scandal when his own establishment was shown to be crawling with vermin. *Bee-Hive* 15 August 1863.
23. John Saville, 'The Baker's Oven', *Arkady Review*, 1955, 18–20, 30–4, 63–8; and Augustus Muir, *The History of Baker Perkins* (Cambridge, 1968), are among the most useful studies of this subject.
24. The best study of the twentieth-century emergence of plant baking is R.F. Banks, 'Labour Relations in the Baking Industry in England and Wales Since 1860; With Special Reference to the Impact of Technical and Economic Change on Union Administration and Bargaining Procedure', PhD Thesis, University of London, 1965.
25. An interesting study of the ease with which men of limited capital could enter the trade may be found in John Blandy's novel *The Tax Earner* (London, 1897).
26. This account is drawn from John Bennett's testimony in the 1862 *Report*, pp.24–6.
27. Home delivery was said to be the way in which some journeymen enhanced their earnings. The tradition of 'making dead men' consisted of charging for extra loaves, and one writer (John Challice, in *The Baker's Friend*, 1862) argued that not one household in ten escaped paying for a half-quatern loaf per week above his actual household consumption. The pamphlet suggested that men whose wages were 18 shillings per week made their places worth 30 shillings by this practice.
28. *1862 Report*, p.26. On the casual work force, see ibid., p.29, the testimony of John Bennett: 'Of the estimated number of 13,000 journeymen bakers in London, many hundreds are continually without regular work. Those without regular work may get one or two days employment at the latter end of the week, or take the places for a time of men who have become sick, of whom there are constantly a great many. The surplus labour in the loaf bread trade in London arises from it being so comparatively easy to learn it. A lad coming from another trade can learn to be a tolerably good hand in a year or two'. See further the observations of Gareth Stedman Jones, *Outcast London: A Study in the Relationship Between Classes in Victorian Society* (Oxford, 1971), p.60.
29. The role of subjective hopes of becoming a small master may have been to make the outlook of some journeymen a conservative one, as is argued in Edwin Dare, 'Thoughts of a Journeyman Baker', *History Workshop* 3 (Spring 1977), 148, which is additionally an interesting description of traditional baking.
30. Sylvia Thrupp, *A Short History of the Worshipful Company of Bakers in London* (n.d., n.p.), pp.7–8.
31. C.R. Fay, 'The Miller and the Baker', *Cambridge Historical Journal* (1923–5), 85–91.

32. John Burnett, 'The History of Food Aduleration in Great Britain in the Nineteenth Century, With Special Reference to Bread, Tea and Beer', PhD Thesis, University of London, 1958, p.24. See the same author's 'The Baking Industry in the Nineteenth Century', *Business History* V, 2 (June 1963), 98–108, for an excellent discussion of the main features of economic change in baking.

33. 1862 *Report*, p.92.

34. Ibid.

35. Ibid., p.93.

36. W.B. Pringle, 'The Regulation of the Price of Bread', *Bakers' Record*, 26 November 1870.

37. J.C. Dwarber, 'The Origin of the Evils in the Baking Trade, Their Treatments, and Proposed Cure', *Bakers Record*, 22 April 1871. For an account of the trial of a master baker convicted of fraudulently obtaining credit from milling companies, see the *Bakers' Record*, 24 July 1869, and letters in the issues of 9 April 1870 and 21 October 1871.

38. Burnett, 'The History of Food Aduleration', *op. cit.*, *passim*.

39. 1862 *Report*, p.39.

40. Pringle, *op. cit.*

41. John Blandy, *The Bakers' Guide* (London, 1886), p.82: 'The steam bakery and "the stores" will so adjust the prices as to leave very little margin for that sort of competition [underselling]. Shall we, who are thus pushed out, complain, though we go down in scores, leaving behind us only the legacy of our white haggard faces, as a haunting memory to those who gave us the last push over the edge and saw us "smash" below? No, we will not complain, but rather bless the hand that has thus stopped the weary struggle; for has it not been clear to us that they have been doing indirectly for many years what they have now done right straight out?'

42. 1862 *Report*, p.25.

43. Ibid., pp.37–8.

44. The percentage of bread baked in underground bakehouses was probably highest in London, and much less in provincial England. Out of 432 bakehouses in Birmingham, for example, only 32 were underground in 1864. It does not appear that underground bakehouses were common in Glasgow, although Carolyn Martyn, in *Trade Unionism, A Lecture Delivered to the Edinburgh Branch of the Bakers' National Federal Union, 9 May 1895* (Edinburgh 1895), refers to underground bakehouses in Edinburgh. the death knell of the underground bakeries was sounded by the Factory Act of 1901, which effectively prevented their spread.

45. 1862 *Report*, p.xxix.

46. Marx, *Capital* (Charles Kerr edn), p.274.

47. 1862 *Report*, pp.106–7.

48. A trade unionist told the Webbs in the 1890s that these distinctions between first, second and third hands in the bakehouses — based not on large differentials of skill, but on seniority — had disappeared.

Sidney and Beatrice Webb, Manuscript Notes on Bakers, Webb Trade Union Collection, British Library of Economics and Political Science, collection E, section A, vol.XLV, 22. (Hereinafter cited as the Webb Collection.)

49. Cf. the excellent analysis by Tremenheere in his *Report* on the question: 'It is . . . strictly in the interests of the men that I have received the evidence that has been tendered to me in the course of this inquiry relative to the imperfection of the Adulteration of Food Act in its present state. If by making its provisions obligatory and more stringent, its action would havethe effect of removing one of the principal causes of night-work followed by many hours of day-work, the result would be a benevolent and beneficial one to many thousands of journeymen.' 1862 *Report*, p.xliii.

50. Gourvish, 'Bread Prices', p.854.

51. [James Ness], *The Incorporation of Bakers of Glasgow* (Glasgow, 1891), pp.19–21.

52. By 1884, of course, very few bakers were using the mill to grind their own wheat. See also J.H. Macadam (ed.), *The Baxter Book of St. Andrew's: A Record of Three Centuries* (Leith, 1903), p.253, for an account of the abolition of exclusive privileges of the incorporations. It would appear, from fragmentary evidence, that the trade in London had definitely lost its guild characteristics by the end of the eighteenth century, while the trade in Glasgow did not completely do so until the 1830s. Scottish economic historiography has not focused much attention on the effects of exclusive trading.

53. 'Glasgow price-setting'; *Glasgow Sentinel*, 18 March 1854; 12 October 1867, 12 October 1872, 24 May 1873; Edinburgh: *Glasgow Sentinel*, 9 June 1855.

54. Glasgow had the largest bakery in the world, which was owned by the United Co-operative Baking Society, at the turn of the century. William Reid, *History of the United Co-operative Baking Society, A Fifty Years' Record 1869–1919*. Glasgow: United Cooperative Baking Society, 1920. See also J.B. Jefferys, *Retail Trading in Britain 1850–1950* (Cambridge, 1954), pp.210–25.

55. C.M. Allan, 'The Genesis of British Urban Redevelopment with Special Reference to Glasgow', *Economic History Review*, 2nd series, XVIII, 3 (1965), 603.

56. See Joy Tivy, 'Population Distribution and Change', in Ronald Miller and Joy Tivy (eds), *The Glasgow Region: A General Survey* (Glasgow, 1958), pp.242–69. Tivy stresses the geographical factors which created this unusual pattern.

57. E.E. Lampard, 'The Urbanising World', in H.J. Dyos and Michael Wolff (eds), *The Victorian City: Images and Realities, I: Past and Present/Numbers of People* (London, 1976), pp.22–3.

58. See the analogous reasoning in William Panschar, *Baking in America*, vol.I, *Economic Development* (Evanston, Ill., 1956), pp.45–54 and *passim*.

59. The distinction between the Scottish manufactories and the small

English bakeries was not that of a mechanised production versus manual labour, but that of a different organisation of work. Rather than a division between first, second and third hands, the Scots were divided between foremen (who earned between 24 and 38 shillings a week) and table hands (who earned 17 to 26 shillings a week), as well as many bound and unbound apprentices. The foremen were members of the union, and were in fact some of its most militant activists (cf. *Glasgow Sentinel*, 9 April 1870). The team system within the bakeries and the consequent rivalry among foremen was said to have been a contributing factor in promoting greater Scottish efficiency. Elsewhere foremen *seem* to have been excluded from the trade union, as was evidenced in the 1872 London strike (*Bakers' Record*, 10 August 1872).

60 Fermentation was produced in Scotland simply by mixing yeast with the flour, chiefly patent yeast, or yeast made by the baker from hops and flour. The London system 'ferments' was absent. London masters thought that the 12-hour system was only possible if one accepted a lower standard of bread; as one remarked, 'It is said that the Scotch make their bread within the 12 hours; but they don't mind how sour the sponge is. The people of London would not eat the common household bread used in some places in Scotland.' 1862 *Report*, p.80.

61. 1862 *Report*, p.28.

62. There is obviously comfort for both the 'leading-sector' and the 'aggregate development' schools of Scottish economic history in this particular case. The durability of corporatist controls in this trade well after their decline in England suggests a far more uneven penetration of capitalism in Scotland, but the demonstrable growth of manufactories in baking after 1846 demonstrates the breadth of changes in production outside the classic industries of textile, coal and metals. See Henry Hamilton, *The Industrial Revolution in Scotland* (Oxford, 1932); S.G.E. Lythe and J. Butt, *An Economic History of Scotland 1100–1939* (Glasgow and London 1975), pp.161–200, among many other titles.

63. The largest London bakery, by contrast, had four ovens and 29 employees, with a few casual workers at the weekend. 1862 *Report*, pp.vii, 32, 37–8. Glowing reports of the Crossmyloof bakery were made by the leaders of the Scottish union, virtually all of whom worked there at one time or another: see the letters in Ebenezer Stevens, *A Voice from A Bake House By An Emancipated White Slave* (London, 1861); *Papers Relating to the Case of the Journeymen Bakers, Published for the Information of the Clergymen, Merchants and Citizens, Who Have Joined in the Requisition to the Lord Mayor of Dublin to Convene a Meeting to Consider Their Case* (Dublin, 1860); and in W. Neilson Hancock, *The Journeymen Bakers' Case* (London, 1862).

64. Strathclyde Regional Archives, TD 66/5/86, TD 66/5/92.

65. 1862 *Report*, p.19.

66. *Glasgow Sentinel*, 25 April 1857.
67. These are the rates cited in *Trades' Societies and Strikes. Report of the Committee on Trades' Societies, Appointed by the National Association for the Promotion of Social Science* (London, 1860; reprinted New York, 1968), p.295.
68. Hobsbawm, *op. cit.*, p.290.
69. An official of the London Union of Operative Bakers remarked: 'with reference to Scotch foremen he thought the best plan would be not to acknowledge them, as very often the Scotch foreman undertakes the work of three men.' *Bakers' Record*, 24 January 1874.
70. *Freeman's Journal*, 23 August 1860; this abolition of night-work did not last.
71. Samuel Ayton reflects on the movement of the 1830s in London in his 'The Autobiography of a Baker, Or, Life as I Have Found It', serialised in the *Bakers' Record* from 6 May 1871 to 19 August 1871.
72. 'Unto the Master Bakers of Edinburgh, The Humble Memorial of the Journeymen at Present in their Employ', reprinted in Stevens, *op. cit.*, pp.31–7.
73. W.A. Guy, *The Case of the Journeymen Bakers: Being a Lecture on the Evils of Night-Work & Long Hours of Labour, Delivered, July 6th, 1848, at the Mechanics' Institute, Southampton Buildings* (London, 2nd edn, 1860), p.25. Gladstone found the proposal to pass a Bill against night-work 'abhorrent to the constitution'.
74. 1862 *Report*, p.18; an anonymous correspondent of the Webbs suggested that the victory was lost in 1847, but this is not clear. It is certain that in 1858 the twelve-hour rule was in force, and most evidence points to the permanency of the system after 1846, although there were local setbacks. Edinburgh bakers seem to have lost the 12-hour system in the 1850s and regained it in 1861 *(Glasgow Sentinel*, 21 December 1861).
75. *Glasgow Sentinel*, 30 November 1867.
76. *Glasgow Sentinel*, 25, 26 April 1857; 2 May, 9 May, 16 May 1957; 5 September 1857.
77. Tom Johnston, *History of the Working Classes in Scotland* (Glasgow, 4th edn, n.d. [1941], p.329.
78. The impact of the campaign is shown by the number of pamphlets and books it produced: W.A. Guy, *The Case of the Journeymen Bakers* (1848, reprinted 1860), *op. cit.;* [A.J. Hunter], *The Results of the Abolition of Sunday Work and Work at Night, in the Baking Trade, in Scotland, Being in Letters from the Secretary of the Journeymen Bakers' National Association of Scotland* (Dublin, n.d. [1860]); John Lilwall, *Bondage in the Bakehouse; or, the Case of the Journeyman Baker* (London, 1860); *Papers Relating to the Journeymen Bakers . . . (op. cit.*, Dublin, 1860); Stevens, *A Voice from A Bake House . . . (op. cit.*, London, 1861); John Jellicoe, *Facts and Fallacies* (Dublin, 1861); W. Neilson Hancock; *The Journeymen*

Bakers' Case (London, 1862); and John Challice, *The Bakers' Friend* (London, 1862).

79. The London Operative Bakers Vigilance Committee undertook this work in London with great effect after 1863. The Bakehouse Regulation Act was wholly repealed by the Factory Act of 1878, which provided for the sanitary arrangement of bakehouses. Under the Factory and Workshops Act of 1883 local authorities were given control over retail bakehouses, leaving the rest under the surveillance of factory inspectors. See C.G.H. Smith, *Handbook on the Law Relating to Bakers and Bakehouses and to the Sale of Bread & Flour Out of London, Containing Introduction, Cases Decided, Notes on Sections, Statutes, &c.* (Coventry, 1893), pp.3–4. Parliament was finally induced to regulate adult male labour in bakehouses — in 1954.

80. For two years John Bennett, former secretary of the Scottish union, lobbied Shaftesbury, Ebury and other reformers, and also Grey the Home Secretary. The Tremenheere Report was commissioned after Shaftesbury championed the bakers' cause in 1861.

81. *Bakers' Record*, 29 April, 6 May, 3 June, 23 December 1871; 27 January, 10 February, 17 February, 11 May, 6, 27 July 1872.

82. *Bee-Hive*, 14 September 1872; 21 September 1872; *Bakers' Record*, 14 September 1872, 28 September 1872, 12 October 1872.

83. Swift, *op. cit.*, pp.258–91; J.J. Barrett, *The Baking Trade in Ireland: Past and Present; or, The Three Great Struggles for Reform in the Condition of the Operative Bakers, 1852, 1859 and 1871* (Maryborough, 1874), pp.59–60; Webb Collection, 65 (comments of Murray Davis, General Secretary of Irish Bakers' Federation).

84. A fuller description of these struggles may be found in I. McKay, 'Trade Unionism in the Baking Industry of Great Britain and Ireland, 1857–1874', MA dissertation, University of Warwick, 1976.

85. Cf. Banks, *op. cit.*, ch.1.

86. The bakers of Dublin managed to preserve as close a control over the labour markets until the mid-1870s, but signally failed to abolish night-work or extend their organisation throughout Ireland.

87. 'When all other classes are obtaining the 54 hours per week, operative bakers should be white slaves no longer, 'cried a letter in the *Bakers' Record*, 27 January 1872. At the same time a speaker was reminding the journeymen bakers of Glasgow that 'throughout the length & breadth of the land other trades had and were progressing, if we did not we ought to be stigmatised as the veriest cowards in existence.' Minute Book of the Journeymans [sic] Bakers Association, Glasgow branch, 16 January 1872. This invaluable document is in the possession of the Scottish Union, at their headquarters in Glasgow.

88. Swift, *op. cit.*, pp.290–1.

89. John Blandy, *The Baker's Catechism* (London, 1891), p.9. Charles Lang told the Glasgow journeymen that they were 'everyday engaged in conducting operations of a highly scientific character,

and must have an education fitted to make them understand what they [were] doing.' *Glasgow Sentinel*, 4 March 1865.

90. This applied with greatest force to the Scottish bakers, whose support for temperance and rational recreation is striking. Their support for extension fo the franchise and their prominence on local trade councils — two features which set them apart from other bakers — are shown in W. Hamish Fraser, 'Trade Unions, Reform and the Election of 1868 in Scotland', *Scottish Historical Review*, L, 2 (October 1971), 138–57. The links between the labour aristocracy and reform are most clearly delineated in Royden Harrison, *Before the Socialists* (London, 1965).

91. 'Unto the Master Bakers', in Stevens, *op. cit.*, pp.290–1.

92. Ibid., p.53.

93. [Hunter], *op. cit.*, p.2.

94. *Bakers' Record*, 24 August 1872.

95. 1862 *Report*, 30.

96. See the reports of John Bennett in the Glasgow Trades Council in the *Glasgow Sentinel*, 21 May 1859; and his comments on suffrage at a meeting of journeymen bakers, *Glasgow Sentinel*, 11 December 1858. When in London Bennett was a supporter of the *Bee-Hive*.

97. As tends to be the case in Trygve R. Tholfsen, *Working-Class Radicalism in Mid-Victorian England* (London, 1976).

98. R.J. Morris, 'Bargaining with Hegemony', *Bulletin of the Society for the Study of Labour History*, 35 (Autumn 1977), 62.

99. Cf. James Dollan, Typescript History of the Scottish Bakers, ch.1, at the headquarters of the Scottish Union of Bakers and Allied Workers, Glasgow.

100. 1862 *Report*, p.31: 'In Scotland,' Bennett claimed, 'since 1846, when the short hours were adopted, the tendency has been to throw the baking trade more and more into the hands of capitalists.'

101. Rules of the Journeymen Bakers' Association, adopted at the meeting of delegates, held at Glasgow, on 17, and 18 August 1858, reprinted in Stevens, *op. cit.*, p.42.

102. See the writings of Samuel Ayton and John Rose in the *Bakers' Record*, 12 March 1870; 16 September 1871; 31 August 1872.

103. The exception would be the local unions of Dublin.

104. *Glasgow Sentinel*, 20 August 1870; *Bakers' Record*, 27 August 1870.

105. Rules, *op. cit.*

106. Ibid.

107. Webb Collection, p.51.

108. W.H. Marwick, *A Short History of Labour in Scotland* (Edinburgh and London, 1967), 37.

109. Robert Wells, *The Modern Practical Baker* (Manchester and London, n.d.), pp.102–3.

110. *Bakers' Record*, 28 March 1874.

111. *Bakers' Record*, 5 February 1870.

112. *Reynolds' Weekly*, 26 May 1850.

113. Webb Collection, p.30.

114. *Trades' Societies and Strikes, op. cit.*, p.296. A.J. Hunter explained that 'the bakers did not seek to limit the number of apprentices, but wished them to be regularly bound, and were very particular as to the admission of men here into their society unless they had served a regular apprenticeship'. *Glasgow Sentinel*, August 2 1862.

115. This analysis of the age structure of the two trades is drawn from the occupational tables of the 1851 and 1871 Censuses.

116. *Glasgow Sentinel*, 12 November 1853 and 21 october 1865; Glasgow Branch Minutes, 16 January 1872.

117. Ibid., 22 August 1871.

118. Carolyn Martyn, *Trade Unionism, A Lecture Delivered to the Edinburgh Branch of the Bakers' National Federal Union*, 9 May 1895 (Edinburgh 1895), pp.10–11.

3. Sailmakers: The Maintenance of the Craft Traditions in the Age of Steam

Mark Hirsch

It is an irony of labour history that the successful organisation of unskilled workers into general unions has been accompanied by significant increases in craft union membership. In Britain, the new unionism of 1889 encouraged many skilled workers to join exclusive craft unions from which they had hitherto stood aloof. In the United States, the conservative craft-oriented American Federation of Labor grew just as rapidly and eventually surpassed the membership levels of the nascent Congress of Industrial Organisations, which initiated the successful drives to organise unskilled mass production workers into industrial unions in the 1930s.[1] Both movements laid bare the class nature of their respective societies and opened new possibilities for skilled and unskilled workers to unite over common class issues.

Apart from the fact that they joined craft unions, little is known about the relationship between skilled craftsmen and mass movements of unskilled workers. Henry Pelling has argued that, as a result of the new unionism, 'the old unionists' attitude to the unskilled began to change', but little evidence is marshalled to indicate precisely which skilled workers' attitudes changed, and to what degree. He notes that 'increasing mechanisation' threatened 'the status of the craftsmen', encouraging them to embrace the unskilled, but again little evidence is brought to bear.[2]

These pages focus upon the responses of one group of skilled craftsmen who were profoundly affected by increasing mechanisation and by the new unionism of 1889 — the sailmakers. Practitioners of a pre-industrial trade characterised by small-shop production, sailmakers figure as archetypal mid-Victorian craftsmen. In their trade policy, sail-makers pursued a programme which approximated that of the New Model unionists.[3] Proud of their skills with a palm and needle, sailmakers subscribed to the first principle of craft unionism — exclusiveness. They were informed by a belief which held that legally-apprenticed men had an inviolable right to participate in making the rules which governed the workplace. This involved determining wages and prices, controlling the labour market by limiting the number of apprentices, and demanding that the quality of craftsmanship be kept at the highest level. And like other craftsmen who believed in the dignity and 'manliness' of skilled labour, sailmakers formed local trade societies to defend the customs of their craft against violations by their employers. Eschewing strikes, they at all times advocated industrial conciliation and arbitration as the best method of settling trade disputes. These beliefs, along with their 'Lib–Lab' political orientation, language and manner of dress suggests that sailmakers might well be labelled 'labour aristocrats'.

In point of fact, sailmakers were neither labour aristocrats nor typical mid-Victorian craftsmen. Unlike the skilled workers who benefited from the expansion and stabilisation of British capitalism in the third quarter of the nineteenth century,[4] sailmakers and other craftsmen in ancillary trades began to experience a sharp deterioration in their living and working conditions. The steady rise of steam shipping not only made it difficult for them to defend their time-honoured craft customs, but eventually threatened to render their craft obsolete.

Craftsmen associated with sailing ships interpreted and responded to this crisis in various ways. Sailmakers re-sponded in 1889 by forming the Federation of Great Britain and Ireland. Organised 'to secure the interests of the sailmaking trade',[5] the Federation was an attempt to preserve the craft and to control the sailmakers' destiny in the face of extreme adversity: increasing steam shipping, changes in the methods of production, and dilution of the labour force with 'non-legal' men and women. A loose federation of old, local friendly societies, the Federation of

Sailmakers was for most of its 32-year existence a small and somewhat trivial organisation — one which the Webbs cynically described as 'archaic'.[6] With a peak membership of approximately 1530, the Federation's main function was to serve as an information centre, sending trade reports to the member branches informing them of the availability of work in each port. Yet, despite the fact that Federation property consisted of little more than a mimeograph machine and stationery; that whole branches folded overnight; that annual conferences were frequently attended by a handful of men — despite this, sailmakers have a claim to a place in the historiography of nineteenth-century craftsmen. They were not the first — nor will they be the last — workers to face obsolescence, and their efforts to preserve the craft tradition and extend it to new types of work in the age of steam shipping merits attention.

Work in the nineteenth-century sailmaking trade was measured by the seasons and by the cyclical nature of the shipbuilding industry. Faced with a work pattern which fluctuated between 'alternate bouts of intense labour and idleness',[7] sailmakers, during periods of dull trade, collected their tools and 'hit the road' in search of work. Tramping was a significant part of their occupational culture, and like other artisans they undertook long journeys in search of employment. Members of the London Sailmakers' Burial Society, for instance, tramped regularly to Liverpool in the 1820s. Others may well have traversed the eastern coast, stopping in Goole, Hull, Hartlepool, Sunderland, South Shields, Leith and perhaps Dundee before returning home.[8]

Unlike other tramping artisans, sailmakers had an alternative to tramp relief. Because the elements took a heavy toll on sails, every merchant sailing ship carried a sailmaker, who was often one of the vessel's chief petty officers.[9] Although this put them in contact with merchant seamen, sailmakers were nevertheless careful to distinguish themselves from mere labourers. Indeed, sailmakers tended to see themselves and other skilled shipbuilding workers, such as wooden shipwrights, riggers and caulkers, as part of an élite.

During peak periods in the shipbuilding yards, sailmakers

practised their ancient craft in large rectangular workshops. Located close to a shipbuilding yard of dock area, sail-lofts occupied the upper storeys of a warehouse, as the massive proportions of sails required large, uninterrupted floor space.

Lofts were generally owned and operated by a master sailmaker, though in some cases sailmakers worked for sailing ship and fishing companies. In the former case, the master was an artisan who had had the good fortune to amass enough capital to set up on his own. The cost of renting a loft, purchasing canvas, rope, twine, work benches and other materials, as well as employing between four to ten craftsmen was not prohibitive, and an aspiring journeyman could set up on his own fairly easily. The social distance between master and journeyman, therefore, was not great: sailmakers passed from one station to another and back again quite frequently.[10]

The mysteries of the craft were jealously guarded by the sailmakers, and were revealed only gradually when a boy began his apprenticeship. According to an eighteenth-century source, an apprentice could be bound at fifteen years of age 'without any particular genius or education' providing he paid a nominal fee to the master.[11] Non-monetary fees were also required. When a boy was bound in Scotland, he was required to provide a bottle of whisky for his mates, and another when he sat down to sew. In England, he was obliged to set up drinks when he first began to sew and then again when he began to rope the edges of the sail.[12]

The binding of a new boy was only one of the traditional rituals which sailmakers practised in the lofts. Like other nineteenth-century artisans, sailmakers maintained certain informal workshop practices which were not listed in their society rule-books.[13] The custom of footing — supplying drinks for the indulgence of one's workmates — was commonplace, and came to the attention of at least one concerned temperance reformer:[14]

> When the oldest apprentice is first employed to take the measure of work, he must mug the journeymen for this honour. The drink at

expiration of loosing is optional. One informant has known it range from 10s. to £5. Marriage is 5s., backed with from 2d. to 6d. Birthday is from 1s. to 2s. 6d. The trade club meets . . . in the public house; men drink 2d. apiece for the use of the room, and the usual results of this are visible.

The significance of these rituals should not be obscured by the temperance reformer's hostility. The sailmakers were not simply drinking alcohol; they were doing so at work in the employer's time. Like other artisans, sailmakers were imbued with pre-industrial concepts of work and time. Task-oriented, not time-oriented, sailmakers believed that their skills conferred upon them the right to set the pace and shape the character of labour at the point of production. Rituals of drinking and conviviality went a long way towards this end.

After the apprentice had satisfied the appetites of his workmates, he began to learn the secrets of the trade. During his first two years he was relegated to the more pedestrian tasks: hauling canvas and delivering the finished sails. The first step in sailmaking — measuring and cutting canvas — was never performed by him, but by the master or his most trusted hand. Canvas was an expensive item, and a master could ill-afford to have it ruined by a clumsy apprentice.

Sails were cut to the specifications designated by a mast-maker if a vessel were first being built, or by a ship owner or captain if it were being repaired. Sails varied in size according to the mast or yard they were set from. Width was determined by the length of the yard, and depth by the height of the mast.[15] With these measurements the master or journeyman with the steadiest hand proceeded to cut the canvas, cloth by cloth with a sharp knife. These were handed over to the journeymen who began the actual process of sailmaking.

The sailmaker sat on a bench called a 'trestle'. In his lap rested the canvas, and to his right, arranged in holes at the end of the bench were his tools. Attached to the trestle with cords were two metal hooks which served as a third hand for holding the canvas in place. Around his left hand, the sailmaker wore a palm — a ring of cowhide with a dented metal plate which held the needle steady as it was pushed through canvas and rope.

Having sewn the strips, the sailmaker began 'tabling' — hemming the outer edges of the sail. Then he strengthened those parts of the canvas which were more likely, under high winds, to chafe against the masts, yards and ropes. With these reinforced, he made holes in the edges of the canvas through which ropes would eventually pass. Then, using a 'fid' — a 10–24 inch cone made of wood or bone — the sailmaker stretched the hole to its proper circumference. Inside the hole he sewed a 'grommet' — a ring which prevented the rope from cutting into the canvas.

The next operation, one which required the greatest skill, strength and patience, was sewing the bolt rope around the edges of the sail. This was the point where, as one expert noted, 'many a well cut sail is spoiled'.[16] Owing to the toughness of the rope, sailmakers used a mallet to push the needle through. When it had been sewn securely, the craftsman spliced another piece of rope to make the 'cringles' — small bows of rope which fit over the bolt rope. The final operation was 'reefing' — making small holes across the middle of the canvas so as to 'reduce the surface of the sail in proportion to the increase of the wind.'[17]

This concluded the work in the loft. The finished sail was then taken to the ship and delivered to the master rigger. The bending of a new sail — fastening it to its yard — was the final operation in the shipbuilding yard, and was another source of ceremony and social ritual. Custom had it that the ship master 'mugged' the men for their labour.[18]

Throughout this process, sailmakers performed hundreds of minute operations which would distinguish the product of their labour from that of non-apprenticed men. Any manual labourer with a modicum of manual dexterity and strength could sew canvas; but only apprenticed sailmakers knew, for instance, that bolt rope needed to be cross-stitched and kept twisted to avoid stretching and tearing. It was this knowledge of the secrets of the craft which encouraged sailmakers to believe that they had a just claim to artisanal status.

Unlike the informal workshop custom of 'footing', sailmakers maintained certain practices which were formalised in their society rule-books and which were adhered to daily in the sail-lofts. These rules were at once defensive and aggressive: they not only defended the craftsmen against their master's encroachments, but also informed the masters about the limits of their own managerial authority. The object of these rules was job control: through them, the men demanded a say over the structure of employment, over wages and prices, and that the dignity and rights of the craft be respected. Whether sailmakers were ultimately successful in controlling their working environment is far less important than the fact that they attempted to control the work process at all. Control, as Carter Goodrich noted, is an elusive and constantly changing phenomenon, and we may profit from his dictum that one should not be preoccupied with measuring 'the greater or less[er] degree . . . of control exercised', but with 'the nature and policy of the union exercising it'.[19]

Local sailmakers' societies were primarily mutual benefit clubs concerned with the disbursement of friendly benefits. But beyond sick pay and death insurance, the societies operated to defend the customs of the craft. As organisations of skilled men, the societies clung to the principle of exclusiveness. The prerequisite for membership was a certificate of a seven-year indenture — proof that a man was trained in the mysteries of the craft and would uphold and respect its traditions. Sailmakers never waivered on this issue.

An equally integral part of the sailmakers value system was the belief that craftsmanship should be kept at the highest level. This made them buck instinctively at any changes or speed-up in the nature or pace of production. Most 'scamping' took place when the journeymen hemmed the edges of the canvas, reducing the overlap from as much as ½–1½ inches, as well as reducing the number of stitches required to sew it down firmly. The craftsmen despised this, not only because it implied a speed-up, but because they believed the dignity of labour rested on the excellence of craftsmanship. 'The harm done to the journeyman,' complained one sailmaker, 'not only in the amount of his work, but in his character as a worker, is too considerable'.[20]

Honourable craftsmen tended to ignore 'scamping' orders.
One sailmaker from North Shields explained that his master
would 'often cry out to give the tabling a long stitch, but we
seldom take any notice and just go on our usual way.'[21]

The sailmakers' effort to control the structure of employment in the sail-lofts encouraged them to limit the number of
apprentices to three per loft. Apprentices were a profitable
source of income for the masters because they were familiar
with the trade, but were not yet paid a craftsman's wage.
Controlling the number of apprentices, however, became
less of a problem in the final quarter of the century because
few boys were willing to be bound to a dying craft. It was
unskilled men and women whom the sailmakers eventually
found difficult to control.

Many employers regarded the craftsmen's control of the
labour market as an encroachment upon their managerial
prerogatives, and sought to break that control in a variety of
ways. Frequently this took the form of a lock-out. Such was
the case in Swansea where in 1865 sailmakers appealed to
public opinion after nine weeks of idleness:

> In the first place, two apprentices . . . are sufficient to do that part of
> the work necessary to be done before it requires to be completed by
> the journeymen. The masters not being satisfied with that they locked
> us out for not working with a man who had not served his time. We
> now appeal to the good sense of the public . . . and the non-prejudiced
> owners and masters of vessels . . . and ask, is this a fair and honourable
> system of doing business?[22]

The outcome of this conflict is not known, but there can be
little doubt that the men remained firm in their resolve to
control the labour market by limiting the number of
apprentices and by excluding non-legal men.

The sailmakers' attempt to control the structure of
employment also encouraged them to maintain various
work-sharing schemes which limited the amount of overtime
a craftsman could put in while another member was out of
work. In most ports, journeymen could work only one night
of overtime a week, and were subject to severe fines for
infringement. One sailmaker explained the logic of this
restriction: 'We do not want to monopolise the work by

overtime when a fellow member is out of employ. Surely in this the masters cannot be losers, for the labour divided between the two costs no more than if done by one'.[23] This suggests that sailmakers' societies, like other unions of working men, exalted mutuality, and in doing so, challenged the very tenets upon which mid-Victorian bourgeois society rested.

In all these ways, sailmakers in their local societies attempted to exercise control over the character of their labour. Their behaviour was legitimised by the belief that the apprehension of a skill endowed the craftsman with certain inalienable workplace rights which existed above and beyond the wages contract. These beliefs were embodied in local society rule-books, and were strictly adhered to. Those who failed to honour these rules incurred the wrath of the entire society.

It is difficult to know how successful the sailmakers were in defending their craft customs, but according to the Webbs, they 'exercised a very powerful influence over the trade'.[24] By 1880, however, that influence was clearly on the wane. In London, Charles Booth found that sailmaking on the Thames had fallen on hard times:

> Sailmakers used to be a large body: fifteen years ago, at least one thousand of them could find employment, but now there are not more than four hundred, and even for this number there is not sufficient work. Autumn used to be busy times, but now all seasons are slack alike . . . employment is so irregular that a man is lucky to get a fortnights job and considers himself very fortunate indeed if he has work that will last for five consecutive weeks.[25]

The decline of shipbuilding on the Thames may mean that the conditions of London sailmakers had deteriorated more markedly than those of their brothers in the outports.[26] Nevertheless, by 1880, sailmakers everywhere were undergoing a painful collective experience. In Grimsby, Hull, South Shields, in all the major ports and shipbuilding centres, sailmakers were encountering a crisis of enormous proportions. The causes of this crisis and the sailmakers' responses to it require examination.

In their voluminous books and short stories, novelists

enamoured with the sea have waxed nostalgic over the passing of the merchant sailing ship. To them, the fully-rigged vessel was a thing of beauty, whose masts and sails lent the waterside an atmosphere of romance and adventure. The world of sail was a paradise, but a serpent in the garden — the steamship — irrevocably destroyed it. the sea had become 'a used-up drudge, wrinkled and defaced by the churned up wakes of brutal propellers, robbed of its vastness, stripped of its beauty, of its mysteries, and of its promise'.[27]

Artistic licence may sanction this view, but it is sorely lacking as a chronicle of the impact of advancing technology. Steam shipping transformed more than just the pace of shipping or the aesthetics of the harbour: it set into motion important changes in the port labourer's work habits and concept of time, while in many cases threatening livelihoods with extinction.[28] In an effort to come to terms with the age of steam shipping, sailmakers, on the initiative of the Hull Sailmakers' Trade and Benevolent Society, formed the Federation of Sailmakers of Great Britain and Ireland.

The impetus for the Federation's formation was not the complete disappearance of the merchant sailing ship. the year 1889 does not represent a watershed in the transition from sail to steam. This was a protracted process which was not completed until well into the twentieth century. Indeed, as Sidney Pollard reminds us, the process was extremely uneven: 'For many years, all steamers carried sail and later all sailing ships had donkey engines'.[29]

Nevertheless, by 1875, trends in shipping and shipbuilding were becoming painfully clear to craftsmen associated with sailing vessels. By 1878, one leading shipbuilding company in Hull had become an exclusive producer of steamships, as had many firms on the Clyde and Tyne. One observer remarked in 1884 that 'no feature of the annual output . . . during recent years has been more remarkable than the great number of full-powered and capacious steamers built for the various trading companies'.[30]

Quantitative data support these observations. The following table gleaned from *Lloyd's Register of Shipping*, demonstrates the national trends in shipbuilding for selected years:

Table 3.1: The Number and tonnage of sail and steam vessels over 100 tons Built in the UK, 1885, 1887–8, 1894, 1898

Year	Type	Number	Tonnage
1885	sail	141	184,308
	steam	241	265,517
1887	sail	61	70,995
	steam	268	406,112
1888	sail	62	77,380
	steam	422	699,613
1894	sail	58	78,389
	steam	469	873,757
1898	sail	6	2,569
	steam	684	1,301,325

As this table indicates, the construction of large sailing vessels declined precipitously, from 141 in 1885 to 6 in 1898. But these figures do not tell the whole story. As the construction of large sailing vessels declined, so too did their operation as cargo carriers. In Hull, large merchant vessels began to disappear as early as the 1860s when shipowners who lost their timber-carrying ships, replaced them with more capacious and economical steamers. One observer found it 'incredible' that the once formidable Hull fleet, in 1878 consisted 'almost entirely of sloops and schooners'. 'Of the larger vessels there are but two ships and a half a dozen barques and about the same number of brigs.'[31]

Despite this dramatic decline, sailing ships did not disappear. The remaining 'windjammers' and the more numerous but smaller schooners continued to carry on a significant portion of British commercial shipping (see Table 3.2).

Table 3.2 shows that sailing vessels clearly retained a foothold in commercial shipping despite the fact that their number and tonnage had declined. Sailing ships continued to discharge their cargoes in Hull well into the 1890s, giving members of the Hull sailmakers Society the opportunity to repair, if not cut, new sails. In 1890, out of a total of

Table 3.2: Number and tonnage of sailers and steamers belonging to the UK, 1850–1900 (excluding fishing vessels)

Year	Sailing ships	
	Number	Tonnage
1850	23,960	3,337,732
1860	24,776	4,134,390
1870	22,475	4,506,318
1880	19,369	3,779,221
1890	13,852	2,907,405
1900	10,573	2,077,655
	Steam ships	
1850	1,178	167,212
1860	1,899	452,352
1870	3,168	1,111,375
1880	5,235	2,720,551
1890	7,381	5,037,666
1900	9,178	7,202,509

Source: *Royal Commission on the Port of London*, (1902), Cd.1151, *Report*, p.25.

2,530,435 tons of cargo discharged, 359,093 tons — or 14 per cent — continued to arrive under sail.[32] With the exception of a few full-rigged ships, brigs and barques, which ironically lingered as suppliers of coal to the steamers overseas coaling stations, most of the remaining sailing ships were either schooners, fishing sloops and barques engaged in the home or coasting trades.[33] These vessels carried fewer sails and did not employ a sea-going sailmaker.

The consequences of the declining construction and operation of large merchant sailing vessels were devastating for sailmakers in the larger commercial ports. Since fewer vessels were being built, the demand for newly-cut sails declined, and the men increasingly found themselves stitching new seams on old canvas. Tramping to sea, formerly the preserve of younger men, became a regularity for all. But even here, work had become scarce. It was under these circumstances of increasing casualisation of labour and deteriorating status that sailmakers decided to knit their local societies together into a national union.

The transition from sail to steam was only one of the

factors which encouraged sailmakers to enhance their power through federation. Another was the new unionism of 1889, a movement which aimed to organise unskilled and previously unorganised workers into general unions. the growing number of mass strikes and the increasing level of class awareness which the movement produced directly influenced the local sailmakers' societies to band together. In 1889, Hull was the scene of very intense class conflict, where both skilled and unskilled workers, particularly waterside labourers and merchant seamen, formed local unions or joined national ones, and generally paralysed the city's commercial life through frequent militant strikes.[34] It is not surprising that sailmakers followed the example of these militant workers who had begun to imbibe the spirit of trade unionism.

Many of the Federation of Sailmakers founding documents reflect the atmosphere of class antagonism that prevailed during this critical juncture in the history of the labour movement. Thus the first words composed by the Hull Sailmakers' Society on 6 November 1889:

> During the past few months public feeling has drawn its attention towards Trades Unionism. All classes of workmen are combining for their common good. Labour has aroused from its lethargy to show its strength to the world and seeks to crush the tyranny of capital. Skilled and unskilled labourers are forming themselves into gigantic amalgamations and Federations for the purpose of protecting themselves and we . . . think this is a proper time to move in some form or another. . . . The time has arrived when we can no longer remain Isolated. We must of necessity unite. We cannot stand in this age of progress when the working men of England are rising and Trades Unionism is so much in the ascendancy.[35]

'The order of the day is federation,' declared President C.W. Peters at the first annual conference:

> It is forced upon us by unscrupulous employers who . . . think we were created for their convenience, and that our only mission in this world is . . . to work for and minister to their comfort and pleasure. They have an idea that we exist only as slaves, to be content with the crumbs which fall from their table.[36]

By 'unscrupulous employers', sailmakers meant not only

their own employers but the Shipping Federation. Organised in 1890 as part of an employers counter-offensive
against the new unions, the Shipping Federation established
offices in all the main British ports and pledged to defend
the principles which underlay the open shop. As a result of
their work aboard ocean-going vessels, sailmakers were
acutely aware of the Shipping Federation's power, and were
optimistic that a better organised working class could defeat
it. President Peters declared in 1890 that:

> The shipowners have formed this combination with the intention of
> crushing the unions of the men. Well . . . they have set themselves a
> task which . . . will require all the brains, energy, skill and even money
> that they possess. We are not afraid of this gigantic combination, we
> rather hail it with pleasure because . . . we shall have every trade in
> connection with shipping . . . united in one Monster Organization.[37]

The aim of this 'Monster Organization' was not to eliminate
the Shipping Federation or any other employers' association. Sailmakers argued that employers had as much right to
organise on their own behalf as did working men. They
asked only that the employers fight 'in a fair and honourable
manner'.[38]

This request for fair play epitomises the sailmakers'
political beliefs. Although their rhetoric evinced a high level
of class awareness, their political aspirations fell short of
socialism. Eschewing direct attacks upon the ownership of
private property, they looked towards trade unions to
defend workers' economic interests. 'Our aim in joining
Trade Unions,' one sailmaker declared, 'is not to exact the
utmost penny of increased wealth accruing to our labour,
but to defend ourselves against the selfish money-grabbing
employer whose god is money and a desire to die a
millionaire'.[39]

Viewing class relations as a struggle between producers
and wealthy, parasitic non-producers, sailmakers, like many
artisans, maintained a vision of a 'just' social order in which
all workers, both skilled and unskilled, would receive the
full product of their labour. In this society, strikes would be
avoided through peaceful collective bargaining. As one
craftsman put it:

Employer and employee must meet together to discuss any question
that may arise between labour and capital . . . with but one object in
view; the settlement of disputes by argument instead of force. Labour
and Capital must go hand in hand. . . . The workers have an
indisputable right to share . . . the wealth created by their labour, and
the sooner employers recognise the fact the better will it be for all
concerned.[40]

The sailmaker's belief in industrial arbitration and conciliation and in the labour theory of value was not a product of
the new unionism. The ideological origins of these beliefs lay
in their artisanal background and in the trade union
movement of the 1850s and 1860s. With this legacy they
fashioned their critique of industrial society as it entered the
last decade of the nineteenth century.

What was new, however, was their belief that working-
class aspirations could only be achieved through independent labour representation. In the 1870s and 1880s, sailmakers believed that workingmen could influence the Liberal
Party to redress labour's grievances. By the late 1880s,
however, the writings of H.H. Champion and other had
begun to shake their faith in the Liberals and, by 1890, the
subject of independent labour representation became a
hotly debated issue.

The sailmakers' revolt against the Liberal Party stemmed
from a knowledge of history. They were aware that working
men had lacked the franchise for many years, and that 'the
measures brought forward in the Commons had all been in
favour . . . of the capitalist and the landgrabber . . . who
. . . appropriated the enormous amount of wealth produced
by the labour of the toiling masses'.[41] But with the extension
of the franchise, working men had the means to alter their
social conditions. After years of support they concluded that
the Liberals would not and could not serve the interests of
the working class. They might 'open bazaars' or contribute
'to some charitable institution', but beyond these they could
expect little. The way forward was 'to fight for the labour
party' — an organisation consisting of working men which
would represent the interests of working men in the House
of Commons. Thus President Peters:

If we are to be honestly represented . . . it is our duty to choose men from our own ranks, men who know what it is to work, and who have experienced the roughs with but a very small share of the pleasures of this life. It must be obvious to all that a man who has struggled by our side for a bare existence will be more likely to understand our position and represent our interests than one who never did a day's work in his life. . . . I would rather be represented by one Ben Tillett than any number of capitalists in creation. We want more Tom Mann's, John Burn's, and J.H. Wilson's . . . to represent us . . . and . . . we as workers deserve flogging if we don't get them.[42]

Although they advocated the election of new unionist leaders, sailmakers clearly saw themselves as 'old unionists'. Special chastisement was reserved for those who would level the distinctions within the working class, and a holier-than-thou attitude was adopted when new unionists criticised the basic tenets of craft unionism. Tom Mann was singled out for having criticised the craft unionist as 'a man with a fossilised intellect either hopelessly apathetic or supporting a policy that plays directly into the hands of the capitalist exploiter'.[43] To this, one sailmaker replied:

We have been twitted with being fossils by the new unionist, but he seems to forget the men, who years ago, when it was illegal to combine, endured persecution, prosecution, and imprisonment, that they might not only better their condition, but with the idea of leaving the world better than they found it.[44]

It was this fundamental belief that the working class should be divided into discernible strata on the basis of skill which imparted the Federation with its 'old unionist' character. Throughout its history, the Federation continued to uphold those practices and attitudes which had been consolidated on the local level earlier in the nineteenth century. An organisation of old, independent societies bound together to protect a threatened artisanal trade, the Federation never allowed new unionism to disrupt the practice of craft exclusiveness. Despite class conscious rhetoric, advocacy of a labour party, and talk of a 'Monster Organization' of all grades of waterfront and shipbuilding labour, the sailmakers never considered the possibility of admitting non-apprenticed men into their ranks. A sailmaker might join hands with unskilled workers in a movement to establish an independent labour party, but he would never allow those same unskilled hands to put a palm and needle to canvas inside a sail loft.

A policy of craft exclusiveness was viable in the years between 1889 and 1892 because some men were still able to deploy a measure of craft bargaining power. Two successful strikes in Hull and Grimsby demonstrated their momentary power. On 24 June 1890, the Hull branch began what the Webbs described as a successful 'general strike' against their employers[45] The men demanded an increase of 3 shillings per week on their weekly wages of 33 shillings. Many fishing companies, ship chandlery firms and sail lofts agreed to the demands, but ten companies employing approximately 42 men remained intransigent. By November, however, the companies submitted, and the Federation as a whole celebrated its first real victory.[46]

Encouraged by the success of the Hull branch, 80 non-union Grimsby sailmakers struck on 18 August for an increase of 3 shillings on their weekly wages of 30 shillings. These men worked almost exclusively for fishing companies, and had chosen a particularly good time to strike; August being the month when fishing smacks were fitted with new winter sails. On 21 August, the Grimsby men announced their intention of joining the Federation of Sailmakers. The companies responded by recruiting blacklegs from the south, and reminded the public that their shops had always been non-union. The Grimsby Trades and Labour Council eventually stepped in to help arbitrate the dispute, eliciting an advance of 1s 6d, as well as a reduction of hours from 54 to 53 a week. The men accepted this offer on the condition that all non-union men be sent back to their home ports.[47]

These strikes stand out as anomalies because they were successful. The increasing number of unemployed craftsmen and the future dilution of the labour force with semi-skilled men and women would seriously diminish their craft bargaining power. Nevertheless, as the policy of craft exclusiveness became harder to defend, the men became correspondingly firm in their resolve to establish it.

The history of the Federation of Sailmakers is one of an agonisingly long search for work, first in sailmaking, and then in any trade which required the stitching of canvas. This latter path was littered with obstacles, for, as they moved out of sailmaking and into other forms of work which required the cutting and stitching of canvas, sailmakers found themselves engaged in various demarcation disputes with other craft unions which claimed similar work.[48] Sailmakers frequently fought with riggers, joiners and other craftsmen for work which required a palm and needle — work which they believed their skills gave them an exclusive right to. In fact, this claim represented the extension and defence of the craft tradition to new areas of employment. Such predatory habits may not have been becoming of skilled craftsmen, but without such behaviour, sailmakers knew they would be rendered completely obsolete.

As they failed to secure work in their traditional craft; as they engaged in demarcation disputes with other craft unions; and as they found it difficult to maintain their craft customs, sailmakers were forced to the realisation that their trade was dying. Some tried to ignore reality by maintaining that the trade was undergoing a temporary slump. But, by and large, most sailmakers by 1897 had begun to speak about sailmaking in the past tense.

Ordinarily, most craftsmen spoke with their feet. Trade reports from the various branches told a similar story: men were working 'clear of the trade', and in doing so, severed their connection to the Federation. 'Sailmaking everywhere is feeling the effects of steam,' confessed F.J. Campbell, the Federation's organising secretary. 'The small amount of work that is to be had at the trade does not promote the feeling of organisation', and as such 'I think the Office of Organising Secretary is of no avail and therefore should cease to exist.'[49]

The triumph of steam over commercial sail was the product of a burgeoning capitalist world economy which required larger and faster vessels to transport goods and raw materials to all corners of the globe.[50] In this age of steam, sailing ships became anachronisms and sailmakers simply became obsolete. But obsolescence was not merely a product of the passing of commercial sail. It was also a part of a conscious policy to substitute machinery for the hands of the skilled craftsman.

Like the transition from sail to steam, the introduction of sewing machines into sailmaking was an uneven process. Machines first appeared in Aberdeen in 1851 and in Glasgow in 1873. Master sailmakers in London first introduced machines in 1877, and the men quickly resolved to restrict their operation to fully apprenticed men.[51] Generally speaking, however, machines did not become a regular part of production until the late 1880s.

The evidence suggests that machines were introduced in order to diversify production. In the early part of the nineteenth century, most sailmakers worked in lofts which were owned by men who had been journeymen themselves, and who supervised the production of a single commodity, sails. As sailing ships disappeared, so too did those firms which produced them exclusively. Master sailmakers realised that to remain exclusive producers of sail was to invite financial disaster, and consequently sought to branch out into new areas of production. Like their employees, they adopted work which was similar to sailmaking, work which required the measuring, cutting and stitching of canvas. The employer that rented or purchased a machine was more likely to succeed in the transition from single to multi-commodity production.

Many Hull and Liverpool sailmaking firms diversified in precisely this way. Once an exclusive producer of sails, Charles Ware's loft had 'kept pace, step by step with the . . . movements of progress', and began manufacturing waterproof wagon covers, coal bags and hatch tarpaulings. Thomas Fussey began producing 'marquees of all kinds . . . furnishing them with chandeliers, ottomans, and rich decorations, laying down first-rate dancing floors, and in fact producing the very perfection of a ball room or dining room under canvas'.[52]

Sewing machines, then, were introduced not so much as a deliberate way of destroying the craft tradition, but rather as a method of expanding sailmaking firms into new areas of production. Still, the machines had a negative impact on the sailmakers' craft customs, for it relieved them of an integral part of their work — stitching the canvas strips together.

Only one process remained in the hands of the sailmaker: sewing the bolt rope around the edges of the sail. Machines could not yet perform this operation because needles were not strong enough to penetrate hard coils of rope.

The anti-machine movement began in 1891, when the Executive Council received a letter from the Sunderland branch complaining that 'machines [were] playing havoc with the men's labour'. After similar complaints, the Executive sent letters to each branch, inquiring into the number of machines in each loft, as well as their impact upon labour. Similar responses were forthcoming: machines were throwing men out of work. An employer in Hartlepool, for instance, had reduced his workforce from fifteen to two after purchasing a machine, and similar reductions obtained in other ports. Colin M'Lachlin of the Glasgow branch spoke for all sailmakers when he declared: 'We must determine in what relation the sailmaking man is to stand to the sailmaking machine'.[53]

The long and ultimately futile debate over strategy began in 1891 with the informal institution of an overtime ban and the promotion of work-sharing schemes. Members agreed that a reduction of hours would enhance the opportunities for unemployed craftsmen to find work, but most branches were not strong enough to prevail upon their employers. And despite the pleas that craftsmen abstain from overtime, many continued to claim extra work.

In 1893, the Federation discussed the possibility of a general strike, but this tactic was not implemented because the men were highly susceptible to blacklegging. In a buyers' market, employers had their choice of unemployed craftsmen, many of whom were former Federation members.[54]

Employers were not able to impose their will effectively in every port. Some larger branches were able to restrict the introduction of machines until later in the century, but even these paid for the privilege. Here, the spectre of machine-induced unemployment was used as a weapon to keep wages below the craft rate. Frazier and Company used this tactic as late as 1913 against the Cardiff branch. After requesting a penny advance, the company replied that if any advance were granted, or if the company 'met them in any way, to meet competition from other lofts using machines in this district, we will have machines ourselves'. At the bargaining table, the company granted the rise, and added that machines would not be introduced for at least one year. The company held to its word: by 1914, Cardiff sailmakers were reduced to machine-minders.[55]

Stripped of their bargaining power, sailmakers opted for the only remaining solution — restriction of machine work to legal men who would be paid a special wage for minding them. The Federation also urged the men to secure special wages for sewing bolt rope to canvas that had been stitched together by machine.[56] Many craftsmen demanded that men be paid 'not less than half that additional to our present rate', but the amount varied between ports.[57] Although the reasons were unclear, machine men in Hull and Liverpool were the most successful in deciding who did machine work and at what price. 'We only allow non-competitive employers the use of sewing machines,' explained one sailmaker in 1914,[58]

> i.e. steamship companies who do their own work. All our other employers are without machines. . . . All machines in Liverpool are worked by our own members. None others are at the job, even the helpers at machines are sailmakers and apprentices are not allowed to work or assist them until they are in their seventh year.

These men were unique in their ability to maintain the customs of their craft. Most branches did not have the muscle to assert control over hiring or the price of labour, and continued to work machines below the craft rate.

By 1900, most employers had rid their lofts of all but two or three of their craftsmen, in preference for women and girls who worked below the craft rate.[59] Here, craftsmen continued to sew rope around machined canvas and performed any other operation which the machine could not. These men were paid at the craft rate because their skills enabled them to deploy a measure of real craft bargaining power. But they were the last of a dying breed.

The cumulative effects of declining trade, sewing machines and dilution had taken its toll on the Federation, and by the turn of the century it had become a small and beaten organisation. Indeed, its very name had become something of a misnomer, for sailmakers had moved into entirely new types of work including the making of awnings, canvas bags, marquees, ship covers, shooting targets, stretchers, tents and wagon covers. The bulk of this work was done by women at sewing machines, but there were certain operations — such as sewing grommets — where the skilled hand of the sailmaker remained indispensable. Because these skills were essential to the proper manufacturing of these articles, some craftsmen were able to bargain effectively with their employers. The reconstruction of the craft tradition within the context of new work opportunities formed the basis of Frank Gilliard's presidential address to the membership in 1912:

> We must not get it into our minds that we are a dying trade and be ready to lie down and die with it. There is now and will be a vast amount of work for sailmakers in the future. That it will be work of a different character seems very probable, but nonetheless we shall be called upon to do it, and it behooves us therefore to keep ourselves strong in unity that we demand and obtain recognition of our claim to trade status.[60]

The demand for tents, stretchers and canvas bags which the first world war produced gave the Federation a new lease on life, but the renaissance was too short-lived. After 32 years of struggle, the Federation of Sailmakers succumbed to the age of steam.

Dr Pelling's claim that the new unionism and increasing mechanisation transformed the attitudes of skilled workers does not hold for sailmakers. Neither of these phenomena appear to have altered their fundamental trade policies. Indeed, they clung more tenaciously to craft exclusiveness in the 1890s.

The impact of the new unionism on the sailmakers' organisational form is clear. By 1889, the declining construction and operation of merchant sailing ships made it difficult and latterly impossible to defend the craft tradition. Facing rising unemployment and deteriorating status, the sailmakers drew upon the militant organising spirit of 1889 and formed a national federation.

If the new unionism had any lasting impact upon the sailmakers, it was on their larger political beliefs. Before 1889, they hoped to pressure the Liberal Party into redressing labour's grievances. The proliferation of new unionist and socialist politics seriously eroded this faith, and encouraged them to support an independent labour party, consisting of men of their 'own stamp'.

Although they advocated the establishment of a labour party which would advance the interests of the working class as a whole, sailmakers nevertheless maintained that the working class was not monolithic, and continued to view it as a social formation with discernible strata based upon skill. This perception continued to inform their trade policy despite important changes in their political outlook.

An exclusive trade policy and an aspiration for independent working-class representation, these seemingly contradictory beliefs were in fact perfectly compatible elements of the sailmakers' ideology. The distinction which they made between political and economic interests made this possible.

The sailmakers' defence of craft exclusiveness rested partly on tradition. Although it is impossible to prove, Federation sailmakers were probably well advanced in years. (The Webbs' use of the word 'archaic' to describe the Federation may well have been a description of the craftsmen). As a declining craft, sailmaking attracted few apprentices, and the men who practised it in the 1890s probably received their political socialisation in the 1850s and 1860s, the 'classic age' of the labour aristocracy. Whatever its precise political implications, this fact suggests that Federation sailmakers saw a fixed gulf existing between themselves and unskilled men. The apprehension of a skill, they believed, was more than a badge of respectibility: it was a sign of a man's integrity, of his 'manliness', a quality which, as Fred Reid has pointed out, nineteenth-century 'trade unionism was held to depend on'.[61] As 'dishonourable' and un-manly' men, unskilled workers could not be expected to take pride in the craft, or to take a firm bearing towards authority; in short, they could not be good trade unionists. These beliefs, current in the 1850s, shaped the sailmakers' thinking in the 1890s, and encouraged their rejection of non-legal men.

Sailmakers, however, were not informed solely by custom. Victims of obsolescence, they made a choice to bar non-legal men for economic reasons: an all-grades policy in the 1890s would have been self-defeating.

While they barred them from the lofts and the union, sailmakers, though condescending, bore no malice towards unskilled workers. As a result of their work aboard ocean-going vessels, and their physical proximity to other waterside industries, sailmakers were acutely aware of the problems facing dockers, seamen and other workers, and advocated the formation of craft unions to represent all grades of labour. It was in this spirit that they favoured independent labour representation. Although sailmakers perceived the working class divided into two groups, skilled and unskilled, they saw no commensurate political divisions. The working class had the same political interests: by electing working men into the House of Commons, labour could abolish poverty by giving producers the full value of their labour. This 'producerist' ideology did not question property rights in any fundamental sense. Nor did it contain an understanding of the relationship between state power and capitalist production. But it did condemn a society which celebrated individualism and the monopolisation of wealth and parasitic non-producers. If sailmakers could not expect the survival of sailmaking, they did look forward to the day when an independent labour party, representing skilled and unskilled labour, could legislate into existence a more just social order which rewarded producers for their labour.

NOTES

All works published in London unless otherwise indicated.
1. On Britain, see Henry Pelling, *A History of British Trade Unionism*, (1971), p.104; For America, see David Brody, 'The Expansion of the American Labor Movement: Institutional Sources of Stimulus and Restraint', in David Brody (ed.), *The American Labor Movement*, (New York: Harper & Row, 1971), p.132.
2. Pelling, *op. cit.*, p.104.

3. Roydon Harrison, *Before the Socialists* (1965), p.10.
4. Eric Hobsbawm, 'The Labour Aristocracy in Nineteenth-Century Britain', in *Labouring Men*, (1974 edn), pp.272–315; Hobsbawm, *The Age of Capital* (New York: Mentor Books, 1979), p.247.
5. Federation of Sailmakers' Papers (henceforth FOS), MSS. 87/1/1/1, *Minute-Book*, 6 November 1889, p.5, Modern Records Centre, Warwick University Library.
6. S. and B. Webb, *The History of Trade Unionism*, (1920 edn), p.430, footnote.
7. E.P. Thompson uses these words to describe the work pace of artisanal labour. 'Time, Work-Discipline, and Industrial Capitalism', *Past and Present* 38 (December 1967), 73.
8. A description of the occupational culture of the tramping artisan can be found in, E.J. Hobsbawm, 'The Tramping Artisan,', in *Labouring Men*, p.34. For the itinerant habits of the London sailmakers, see, Webb Trade Union Collection, (henceforth WTUC), section A, vol.XXXIII, p.310. British Library of Economic and Political Science, The London School of Economics.
9. Basil Greenhill and Ann Giffard, *The Merchant Sailing Ship*, (Newton Abbot, 1970), p.30; Basil Lubbock, *The Last of the Wind-Jammers*, vol.I (Glasgow, 1927), p.60; William Mountaine, *The Seaman's Vade-Mecum* (1747).
10. This frequent upward and downward mobility is evident in the special provisions in the society rule-books. Usually a journeyman turned master could remain a recipient of the Society's burial fund, but he could not attend meetings. See for example, FOS MSS. 87/5/1, Hull Sailmakers' Trade and Benevolent Society, *Rules*, (1881).
11. R. Campbell, *The London Tradesman*, (1747), p.300.
12. John Dunlop, *The Philosophy of Artificial and Compulsory Drinking Usages*, (1839), pp.11, 188.
13. E.J. Hobsbawm, *Primitive Rebels*, (1959), p.154; Brian Harrison, *Drink and the Victorians*, (1971), p.40.
14. Dunlop, *op. cit.*, p.11.
15. David Steel, *The Art of Sailmaking*, (1809), p.16.
16. Robert Kipping, *The Elements of Sailmaking*, (1847), p.58.
17. Ibid.
18. Dunlop, *op. cit.*, p.11.
19. Carter Goodrich, The Frontier of Control (1975, 1st edn 1920), p.260.
20. FOS MSS. 87/1/4/1 (1892), p.36.
21. FOS MSS. 87/3/3/1, Fythe to Gotham, 1 October 1892.
22. 'The Swansea Sailmakers Lock-Out', *The Bee-Hive*, 22 April 1865.
23. Ibid.
24. WTUC, section A, vol.XXXIII, p.309.
25. Charles Booth, *Life and Labour of the People of London*, (1902), 2nd series, vol.I, *Industry*, pp.278–9.
26. Sidney Pollard, 'The Decline of Shipbuilding on the Thames',

Economic History Review, 2nd ser., III, I (1950–1).

27. Joseph Conrad, *An Outcast of the Islands* (1949), pp.12–13.
28. Ben Tillet was acutely aware of the impact of steamshipping on the dockers' work habits. Work became less cyclical, though by no means less casual, because dockers no longer had to wait for the wind to bring vessels into the harbor. See, 'The Docker: His Life and Work' (n.d.), in the Tillett–McKay Papers, MSS 74/7/6/6/1, Modern Records Centre, Warwick University Library.
29. Sidney Pollard, 'Comment', on Harley, 'The Shift from Sailing Ships to Steamships, 1850–1890', in D. McCloskey (ed.), *Essays on a Mature Economy: Britain after 1840*, (1971), p.135.
30. *The Trade and Commerce of Hull* (Hull, 1878), Appendix ix; David Pollack, *Modern Shipbuilding and the Men Engaged in It* (1884), p.198.
31. *The Trade and Commerce of Hull*, p.101.
32. *Royal Commission on the Port of London* (1902), (Cd.1153), Appendices to the *Minutes of Evidence*, pp.850–2.
33. Schooners continued to bring fruit from the Azores until the second world war. Sail also lingered as the main motive power of barges which brought gain from Ipswich, bricks from Kent creeks, and coal from Yorkshire to London as late as the 1930s. Fishing 'smacks' such as Great Yarmouth's herring 'dandies', Lowestoft's 'shrimpers', and Whitstable's oyster 'yawls' remained motorless until later. See, Basil Greenhill, *The Merchant Schooners*, vol. I (Plymouth 1968), p.19; Robert Simper, *East Coast Sail* (Newton Abbot, 1972), pp.7, 16, 41–54; George Foulser, *Seaman's Voice*, (1961), p.11.
34. Labour militancy in Hull is analysed by Raymond Brown, 'Waterfront Organisation in Hull, 1870–1900', University of Hull, Occasional Papers in Economic and Social History, No.5 (Hull, 1972).
35. FOS, MSS. 87/1/1/1, *Minute Book*, 6 November 1889, pp.2–5.
36. WTUC, section D, no. 251, *Conference Minutes* (1890), p.12.
37. Ibid., p.12.
38. Ibid., *Conference Minutes* (1891), p.9.
39. Ibid., p.7.
40. Ibid., *Conference Minutes* (1890), p.13.
41. Ibid., p.17.
42. Ibid., *Conference Minutes* (1891), p.15.
43. Quoted in E.J. Hobsbawm (ed.), *Labour's Turning Point* (1949), p.72.
44. WTUC, *Conference Minutes* (1890), pp.14–15.
45. WTUC, section A., vol.XXXIII, p.287.
46. *Hull Daily Mail*, 30 July, and 4, 7, 13 August 1890.
47. Ibid., 5, 19 September 1890; *Grimsby Express*, 18, 19 August 1890.
48. Demarcation disputes were common amongst shipbuilding workers on the north-eastern coast in the early 1890s. The roots of these disputes lay in the introduction of new passenger liners, refrigerator ships, and oil tankers which created new types of work for a multitude of different craftsmen. As a result, each was compelled to

fight the other for work which their overlapping abilities and skills entitled them to. See H.A. Clegg *et. al.*, *A History of British Trade Unions Since 1889*, (Oxford, 1964), p.128; P.L.Robertson, 'Demarcation Disputes in British Shipbuilding to 1914', *International Review of Social History*, XX (1975).

49. FOS, MSS. 87/1/4/3, *Conference Minutes* (1897), p.17.
50. E.J. Hobsbawm, *The Age of Capital*, (New York, 1979), p.59.
51. WTUC, section A., vol.XXXIII, p.359.
52. *Industries of Yorkshire*, pt.II, pp.194, 201.
53. FOS, MSS. 87/1/1/1, *Minute Book*, p.126, excerpt of letter from the Sunderland branch, 28 February 1891; FOS, MSS. 87/1/4/1, *Conference Minutes*, 1892, p.26; WTUC, section D, no.253, *Conference Minutes (1893), p.10*.
54. *Ibid. Conference Minutes* (1892), pp.13, 26; FOS, MSS. 87/1/4/2, *Conference Minutes* (1894), p.13.
55. Cardiff Sailmakers' Society Papers, vol.I, MSS. 4/11/66, letter in the account book dated 22 November 1913; letter in the correspondence file date 19 December 1913. Manuscript Room, Cardiff Central Library.
56. WTUC, section D, no.253, *Conference Minutes* (1893), p.31.
57. FOS, MSS. 87/1/1/4, *Minute Book*, 12 September 1898.
58. FOS, MSS. 87/3/9/11, Swift to Hicks, 4 January 1914.
59. FOS, MSS. 87/1/1/4, *Minute Book*, 12 September 1898.
60. FOS, MSS. 87/1/4/7, *Conference Minutes* (1912), p.11.
61. Fred Reid, 'Keir Hardie's Conversion to Socialism', in A. Briggs and J. Saville, (eds), *Essays in Labour History*, vol.II (1972), p.29.

4. The Stamp of Futility: The Staffordshire Potters 1880–1905

Richard Whipp

In 1901 a reviewer in the *Manchester Guardian* remarked of the potters unions:

> The wonder is how a body of workers of singular skill, holding virtual monopoly of a great trade, in what is virtually a single town, should have failed so signally to do for themselves what artisans under more exacting conditions have generally managed to achieve. The stamp of futility is set on everything these unfortunate potters touch.[1]

This study seeks to explain why the trade unionism of the Staffordshire potters was so beset by difficulties and to investigate the central role of the craft unions. The complex and subdivided industry produced a hierarchically-ordered and segmented workforce. Differences among the potters based on the division of labour and attempts at craft control in the face of technological change were the main causes of the potters' difficulties. The antiquated, highly dangerous industry dominated a compact, isolated community and presented a network of problems to the potters which they never fully solved. From 1880, the pottery industry was one of the most important employers of female labour in Britain. It therefore provides a strong example of the relations between men and women as workers and trade unionists in the late nineteenth century.

The Staffordshire Potteries is the largest concentration of ceramic industry in the world. In 1891 the National Order of Potters could justly call the Potteries 'the greatest centre of the trade in the kingdom'. Three-quarters of the country's pottery workers lived within five miles of Stoke Town Hall and they produced 90 per cent of the pottery manufactured in Britain in 1901. Based on a unique combination of productive local coal measures and clays, pottery manufacture around Stoke had developed from its cottage industry base of the early seventeenth century to become a factory industry of national importance by 1800. £4 million worth of pottery was produced annually in 1896 rising to £7¼ million per annum by 1907. In 1901 the industry employed 46,000 workers of whom 21,000 were women, making it the sixth largest employer of female labour in Britain.[3]

The industry dominated both the labour market and environment of the Six Towns. There were coal and iron concerns in the area, but these were dwarfed by the industry which they largely served. In 1898 a group of women potters described their district as 'small in area, compact and comprising practically no other trade'. The Census reports show more than half the population of a quarter of a million directly or indirectly engaged in the staple industry. To an observer in the 1890s the Six Towns were 'to all intents and purposes one place and quite indistinguishable given the industry which united them'. The townscape was composed predominantly of shard rucks (piles of broken earthenware), marl (clay and lime) holes and 2500 ovens. As an industrial region it was isolated from the rest of the country by a rudimentary communications system and the insularity produced by the dominance of the long-established staple industry. There were few reasons to travel to the Potteries in the nineteenth century unless you were involved in pottery.[4]

The pottery industry of the late nineteenth century was subdivided into six main branches. Earthenware was by far the largest with firms throughout the Potteries. By contrast, the china branch was centred on Longton and was almost a self-contained industry. Burslem was the main location of the twenty firms who made the cheaper jet and Rockingham ware. Hanley and Burslem had an even smaller number of firms producing electrical fittings and china furniture. The sanitary trade was highly specialised and had developed only since the 1870s in response to the increasing demand from public utilities and domestic house-building. Tile manufacturers accounted for one-eighth of all pottery firms. Manufacturers often produced ware for more than one branch of the industry. Many, like Minton's, combined earthenware and china production.

Each branch of the industry was further subdivided according to the market and product type the component firms catered for. In this respect Frederick Parkin considered even the small jet and Rockingham trade 'presented something like a jig-saw puzzle to the observer'. The range of firms found in just one branch of the industry was extensive. The china trade contained Copeland's and Wedgwood who produced porcelain for international markets, as well as the 'penny-jack' shops in the back streets of Longton which survived on work subcontracted from larger firms. As one potter concluded in 1891: 'when an effort is made to treat as a whole such a complicated and varied calling as the potter's it is next to impossible'.[5]

The market conditions of the late nineteenth-century pottery industry provided the main context for the attitudes and actions of the manufacturers. The producing power of some potteries had effectively doubled after the introduction of relatively simple machinery. Demand did not match the increased supply. In the 1890s particularly, demand contracted in the domestic market. This led one employer to remark 'anybody can make pots now, the only art about it is how to sell 'em'. Competition intensified and the familiar technique of price-cutting accompanied it. Prices were said to be at the lowest ever known from 1892, leading to record numbers of bankruptcies. Costs were cut as owners struggled to maintain profit levels since most firms' capital came from retained profits. *The Gazette* published a series of articles dealing with the problem of the moment. In one it asked, 'What are the most approved methods of lowering wages and otherwise reducing the costs of production?' Besides lowering potters' wages, such intense competition made the prospect of collective action by manufacturers even less likely.[6]

Though exports were rising in the 1880s, 1890–1905 were poor trading years for the industry. From 1870 to 1889 exports had risen in value from £1.8 million to £2.3 million per annum. By 1894 they had fallen back to their 1870 level, where they remained until 1902. The McKinley and Dingley tariffs in America, and the increased production of cheap foreign ware, threw more pottery on an already depressed domestic market. As the comprehensive US report on world pottery production noted in 1915, this led to Staffordshire pottery 'being sold at a price which was not fairly remunerative . . . with the result that wages were gradually brought down to a very unsatisfactory level'. The Board of Trade and trade union reports consequently show that from 1890 to 1902 almost every sector of the industry was in some way adversely affected by the poor trade. Unemployment was common to all levels of pottery worker during this period. Some had short phases of prosperity, such as the printers based upon short-term demand for specialist printed ware in Australia. However, most potters, even sanitary pressers and ovenmen, could not escape the effects of the adverse market conditions.[7]

The wide range of firm type in the industry hindered the development of a broader industrial unionism whilst at the same time affecting the power of the existing unions. Until the late 1920s the industry was composed almost entirely of private companies. These ranged from the operations of Brown–Westhead, Moore with 700 employees, to the 10-man workshop rented for 5 shillings a week. The predominant form of business organisation was the family firm. This supplied adequate capital for the small amount of plant needed in potting. The abundant local labour supply enabled firms to rely on intensive use of labour. The management techniques required were modest as long as the industry was run on a predominantly craft basis. This level of expertise the family firm could adequately supply. Combines or multi-plant operations were almost unheard of; individualism remained the distinctive feature of most pottery firms.[8]

The large number of small firms posed difficulties for the potter, unionist and non-unionist alike. Small-sized firms were a feature of the industry until 1945. As late as 1939, 221 out of the industry's 327 firms employed less than 200 workers each. With bankruptcies at an all-time high during the 1890s (Davenports ceased trading in 1892, and Adams in 1893) small *ad hoc* firms were set up on the hope of slender but quick profits. One potter thought the main feature of the period had been 'the decay of big firms and the uprising of many small ones . . . as men with little capital (and not much more experience) have rushed into the business'. Competition amongst the smaller potbanks, producing the cheaper ware, was described as 'suicidal' in 1890. Price-cutting by these firms in order to gain a share in a contracting market led to nearly two decades of 'slaughter shop' prices. The pricing agreement, suggested by Owen in 1890, was rejected by the manufacturers as unworkable, since they were unable to guarantee the compliance of the smaller producers. A disillusioned ovenman therefore wrote in 1890:

> One of the evils of the trade is the facilities now offered for outsiders and greedy members of the trade to enter it as masters . . . If the more legitimate employers would co-operate with the better class of workmen then something could be done to rescue the potting from the harpies . . . as one said to me lately, in his vulgar way, 'I intends to buy my labour at as low a price as I can get!'

With prices and costs being cut in general, wages were pared down to a minimum. Trade unionism was seriously weakened by the low wages prevailing in the smaller potbanks. Longton's china trade in this respect stood out. Noah Parkes of the printers and transferrers and Thomas Pickin for the hollow-ware pressers both found the 'breakneck competition' of the masters and resultant low wages made Longton 'notoriously bad for trade unionism'.[9]

Scattered among the six towns, the 400 potbanks presented a formidable organisational problem. Mary MacArthur and her WTUL organisers found 'the large area the Potteries covered made it physically difficult to attempt to visit them all'. Even the larger craft unions faced problems of organisation in an industry dispersed among so many small, scattered units. The Printers and Transferrers, with the 'indefatigable' George Ingleby, were unable to develop a comprehensive system which could effectively deal with individual disputes in the myriad of workshops. At best the craft unions could only give guidelines to their members for their own individual bargaining it being impossible to represent every member in every last workshop. The notice put out by the printers and transferrers executive after the

1900 strike is illustration of their difficulties. It asked 'that the printers and transferrers who make individual settlements are earnestly requested to see that they settle on the full terms as set forth in the appeal, and where possible the agreement should be ratified by the officials of the Society'.[10]

In the larger works, combination was retarded by the dependence of workers on their firm's internal labour market. Works such as Wedgwoods and Doultons had between 50 and 100 departments. The potter was tied to his firm in two main ways. First, each potbank developed its own highly specific clay recipes and production techniques. These were developed empirically and shrouded in secrecy. Thus managers were loathe to allow skilled potters to move freely to other firms and workers themselves found it difficult to learn the new methods of another company. Secondly, semi-skilled workers especially were influenced by internal job ladders. For example, designers were often promoted from the mould-makers, modellers or painters, each of whom would have already served a seven-year apprenticeship. Before becoming a craftsman turner a potter would have progressed from odd-boy to mould-runner, to attendant and thence to apprentice. Though under attack in the smaller potbanks, apprenticeships were usually enforced by the larger firms, as the indenture cases at Burslem stipendiary court show. Each department of a potbank had a headman or headwoman promoted from the more experienced potters. It was not unusual for a potter to remain with one firm as he or she worked up the promotional scale. Elsie Grocott was employed by Firkhams (and later by Portmeirion when they took over Kirkhams) throughout her 40 years of working life. Long-established firms employed successive generations of potter's families, and loyalty was highly prized and rewarded, leaving trade unionism surrounded with suspicion and blunting its appeal.[11]

Employer paternalism accentuated the dependence of the potter on his firm and made union organising all the more difficult. Many of these firms were private, family concerns still owned and run by a master potter and his relatives. For some this produced 'a cosy sort of industry' with 'a close personal link between the operative and employer'. The MacIntyre family held an annual dinner for its staff. William Ward's retirement in 1879 as foreman at Wedgwood's was marked by a feast for the whole company. Such treatment was certainly popular with some workers; one wrote to the *Potter's Examiner*: 'It is by showing that they regard their workpeople as honourable co-workers that such great firms as Wedgwood's "call forth the Monday" at Etruria Inn.'

The larger owners developed their paternalism on a grander scale. The free library at Tunstall, the public parks, the Wedgwood and Minton institutes and the annual Meakin concerts are the largest examples. Such expenditure was repeatedly cited during the course of disputes by manufacturers as examples of their beneficence. To some it appeared a creditable argument against the demands of unionists and deprived the latter of support.[12]

The types of treatment that the potters' unions received from manufacturers was varied, but made their task no easier. Copelands and Mintons for example declined to implement the Hatherton wage reduction. Other manufacturers were 'kind and considerate throughout'. One unionist stated publicly, 'Let honour be given where it is due. Union is not intended to harass those who righteously recognise labour's claims, but to build them up in their position'. The Brownfield pottery openly advised its female workers to join the newly-formed WTUL branch in 1892. Broadly speaking, the larger, more modern and efficient firms seem to have had the best conditions and wages. However a large number of manufacturers felt a clear antipathy towards trade unions and others seriously misunderstood the basic aspirations of the unionist. Their reactions to the potters' unions ranged from the banning of trade unionists from their premises to the deliberate precipitation by some of the 1890 dispute in order to break the main unions. Some openly condemned the union as 'nothing more than a congregation of machines'. Others resented having to bargain with union officials instead of the individual bargaining between employer and operative. One manufacturer considered the unions were 'interfering generally with that freedom and independence which had hitherto been so highly prized between employer and employed'. Certain owners were less subtle in their objections. 'Picketing is simply a terrorism', and 'the unions are at the bottom of all trade troubles' thought F.R. Benham, chairman of the North Staffordshire Pottery Manufacturers in 1906.[13]

Women and unorganised potters employed by the latter type of manufacturer were in a difficult position. Mary Marland discovered how 'the doors of many of them are kept vigorously closed against all but the workers . . . and to whom it is hardly possible to convey any words of advice and encouragement . . . they have so much to bear, but they fear that to say anything would make things worse'. Some owners were openly hostile to women's trade unions. As Evelyn Mach-Phillips admitted, many a woman potter would not join a union since 'her dismissal was a foregone conclusion'. In attempting to reduce costs, manufacturers spoke plainly of using women as cheap labour and as a means of breaking down craft opposition to new techniques. The manufacturers' spokesman at the 1891 arbitration responded to the cup and saucermakers' demands thus: 'if they don't care to take our terms we can supply their places with women and apprentices'.

The poor collective organisation of the pottery employers hindered the attempts of potters to combine at every level. A divided workforce and a diffuse collection of manufacturers produced the haphazard bargaining patterns within the industry. Collective bargaining was so difficult even within one branch of the industry. Instead, bargaining took place within the potbanks, if not the workshop, between employer and potter. The trade journal repeatedly bemoaned this lack of organisation or coherent action among the manufacturers. In 1894 there were more firms outside the manufacturers' association than there were inside. By 1906 their association was still small with those outside 'simply ignoring it'. Industry-wide agreements were therefore difficult to produce or enforce. The sectional awards of the 1900 wage negotiation were ignored by many firms. Similarly, the disunity amongst the employers was the main cause of the failure of the arbitration and conciliation board after 1892. As a result, each union was forced back to sectional craft bargaining and the negotiating of agreements plant by plant. Burton, as chairman of the manufacturers' association, complained in 1908 that the action of his committee had 'been very much hampered . . . by the failure of so many manufacturers to attend to it at all, or to take any notice of its doings or recommendations'.[14]

Any account of the industrial and labour process of potting must recognise the almost limitless complexities involved. Only the main sequence of events will be dealt with here. Excluding tile-making, the main stages of manufacture are relatively uniform to each branch of the industry. However the complex processes necessitated a large variety of occupations and levels of skill with the largest firms dividing their production process into 100 separate departments by 1900. The following is an outline therefore of the crafts and occupations of the potters.[15]

The production process began in the slip house. The necessary raw materials were ground in a blunger and mixed with water to form a slip of liquid clay. Excess water in the slip was evaporated off in a slip kiln or, towards the end of the century, via a clay press.[16] The clay was then wedged by hand or worked in a pug mill to achieve a uniform density and consistency that rendered it plastic and workable. In the slip house the head slipman was responsible for the production of the slip. Though small in number, the head slipmen held a position of considerable importance based upon their skill and responsibility. Not only did each recipe have to be meticulously adhered to but the differing qualities of plasticity (as required by plate-makers as opposed to hollow-ware pressers) had to be finely judged. Serious loss of wages for other potters would result if ware was spoilt by air bubbles or uneven contraction due to faulty mixing. The headman would be assisted by two or three adult males, largely unskilled, to tend the blunging, filtering and pugging. Below them women or male youths would perform the arduous task of hand wedging or carrying the clay to and from the cellars. Wedging involved raising a slab of clay above head height and bringing it down with great force on a bench in order to expel any air bubbles. In 1903 one woman was recorded wedging 32 slabs of clay per day, each weighting ½ cwt.[17]

The second stage of production took place in the potting shops. Here the clay was manipulated by hand or machine to produce the required shape. Four main crafts were involved. The thrower received a ball of clay which he 'threw' on a wheel using the combination of the wheel's rotary movement and his manual skill. After drying, the ware was then turned on a lathe by a turner who scoured out decorations and 'finished' the surface of the article. There were two kinds of presser. The flat presser placed a plaster mould on the head of a jigger which shaped the upper surface of the article. The back of the article was produced by the application of a profile. The second type, the hollow-ware presser, shaped hollow containers such as jugs or ewers. He used two halves of a mould, into which he would press a 'bat' (a beaten lump of clay), trim the edges and finally join the two halves together. Once the plaster mould had absorbed the moisture from the clay the mould would be dismantled and the article removed.

All the throwers, turners and hollow-ware pressers were male. Most flat-pressers had traditionally been male but by 1890 an estimated 1000 journeywomen flat-pressers were working on the smaller, cheaper flat ware. Each of these four types of craft workers would sub-employ from one to four children or women attendants. The attendant's job was to carry clay, run the moulds and beat out the clay ready for pressing. If steam power were not available, women were used to turn the potter's wheel by hand and to work a foot treadle to power the turner's lathe. Women and youths would also be sub-employed as fettlers, towers and spongers who produced a smooth finish to the ware. All these sub-employees were classed as unskilled.[18]

In the third stage of production, the firing, a separate hierarchy of jobs existed. The ware had to be fired in two types of oven. The biscuit firing occurred while the ware was still in its dry or biscuit state; the second, the glost firing took place after the ware had been glazed. The fireman's position was one of the most important on the potbank. He was in complete control of the ware produced by all the preceding stages of production. Any miscalculation on his part during the 50 hours it took to fire the 20-foot-high glost ovens, or the 45–60 hours for the biscuit oven, and thousands of pounds worth of ware could be ruined. Below him the chief ovenman employed and supervised the gangs who filled and emptied the ovens with saggars (the protective marl clay vessels which protected the ware during firing). It was a large operation. One oven could take 2350 saggars, each weighing 1 cwt. A total of 58 tons had to be handled twice in two days. Women, usually as casual labour, were sub-employed to empty the saggars and sometimes 'place' them in the china furniture and electrical fittings branch.[19]

Before the glost firing, the ware had to be dipped in glaze. This was carried out in the fourth sub-department, the dipping house. Here male and female dippers dipped the ware in tubs of glaze, a job requiring varying amounts of skill to produce an even coating of glaze. Dippers too would sub-employ assistants to 'put-up' and 'take-off' the ware and 'lawn' or filter the glaze with muslin cloth. Within the dipping house groups of women and children, employed casually, would clean and finish the ware after each firing, and clean and sort the protective stilts and spurs.

The fifth major department was the printing shop. The male printer used a copper plate engraving to print a pattern onto a transfer sheet. This pattern was then applied to the biscuit ware by a journeywomen transferrer. Both operations demanded high levels of skill. An apprentice would usually rub the paper the final time on the ware and wash off any surplus paper. A girl cutter's job was to divide the patterns into separate designs for each article. The printing shop also produced its own sub-employment system. The printer sub-employed the transferrer who in turn paid the apprentice and cutter.[20]

The decorating department of most potbanks contained a wide range of occupations, the bulk of them performed by women. These ranged from professional artists to young girls dabbing colour on cheap ornaments. Painters varied with regard to levels of skill and only a few served formal apprenticeships. Other types of decorators included the ground-layers and the gilders. The former painted a pattern on the ware with an adhesive oil upon which dry powdered colour was then dusted, or in some cases aerographed. Gilders dusted powdered gold onto ware which burnishers then polished. Finally, in the last department the warehouse, a head warehouseman supervised the the final polishing, sorting and ordering of the ware. Even the male packers considered themselves a separate sub-department at the end of the production process, as they constructed the wooden packing cases and then packed them with completed ware.

Given that these seven departments were only the major stages of the industrial process the division of labour in any one potbank was extremely intricate. Obstacles to collective action were in-built. Departments worked entirely separately from each other. Most potbanks were composed of a large number of individual workshops. The various shops were considered to be 'places by themselves, with hardly any supervision', by H.M.I. Redgrave and 'in that respect rather an extraordinary trade . . . [as] the ordinary discipline of a factory or machine shop is not brought to bear upon them'. Different departments had different interests. Delays or mistakes in the production process produced loss of income and often provoked anger between departments. For example, a bad pattern on a tile could be caused by a number of departments. It could be the fault of the colour-mixer or the glaze-maker, the tile may not have been placed level in the kiln, or the painting may have been faulty. It was an almost impossible problem to solve and in the absence of formal methods of determining the cause the weakest sections received the blame. An observer of such disputes in 1906 found that 'the unfortunate person who is generally supposed to be responsible for it is the female, the paintress, because she has the least protection . . . of the others in the works, the men can defend themselves'.[21] Joint action found it difficult to flourish where few joint interests existed.

The social relations of the potbanks were founded on the basic divisions apparent within the workforce. A major difference between male and female potters was that of age. Clara Collet in 1892 found that 72.6 per cent of women potters in her study were employed between the ages of 13 and 25. However the 1901 Census shows that 76.3 per cent of the men were at work from 20 to 55 years of age. One of the industry's most notable ventures was the employment of juvenile labour. In the 13–20 age group girls outnumbered boys by almost 2:1. Being much younger than her male sub-employer or workmates, and leading a much shorter working life, these factors helped to ensure that the female potter occupied the lower positions in the potbanks in terms of status and authority.[22]

The second main division was created by the system of family employment. Since the seventeenth century male craftsmen had been paid to produce a count of ware at a given price. He supplied his own assistants who took the form of his own family or kin. Despite the development of factory industry, and of formalised relationships within the potbanks, craftsmen continued to sub-employ their own families. William Owen informed the Royal Commission on Labour that it had been 'the custom from time immemorial', and saw no reason for it to change. With low wage levels women and girls were often compelled to find work in potting outside their family work group, in order to make up deficiencies in the family income. In 1904 Hilda Martindale discovered that in the Potteries, 'a woman is looked upon as lazy unless she takes her share in contributing to the family income'. Adelaide Anderson also thought that in nine cases out of ten women went to work in the potbanks due to 'strong domestic or economic pressure'. Women potters, unlike their male counterparts, were not regarded as working for themselves but as subordinates within a family work group or only in order to make up a family income when necessary.[23]

The division of labour in the pottery industry, based on sub-employment, was the most divisive force among the potters. Craftsmen were seen to assume 'all the traits of an employer'. The hollow-ware pressers' use of cheap female labour was copied by many pottery owners in the 1880s and 1890s. The immediate source of authority over most women was their male sub-employer. In 1880 it was 'an almost universal custom in the trade for the printer to be held responsible for the attendance and supply of his female assistants'. A girl assistant in 1892 described her situation thus:' the girls have nothing to do with their employers and if anyone complained to the employers of the man under whom she worked the man would pay her out for doing so'. Sub-employers could alone determine the wage, conditions and duration of their sub-employees' work. Women, too, sub-employed girl assistants as in the case of the dippers or transferrers. The system led Sylvia Pankhurst to describe the potter as 'the slave of a slave'.[24]

The sub-employment system inhibited combination in two ways. First, it broke the potters down into even smaller, separate work groups. Within such groups the members' aspirations were not alike. The 5 shilling a week mould-runner or batter and ballers had little in common with their presser employer earning 30 shillings. Secondly, the system produced serious problems specific to the female potter. The woman assistant as a member of a family work group, or helping to contribute to a family income, was in a difficult position. While she might see her position as subordinate, to act to change it would involve the straining of family ties. Low wages and her insecure employment made the potential for independent action among single women equally poor. Receiving wages deemed only to make up a family income, the first aim of most single women was subsistence. Singly employed in small work groups women had few other women on hand with whom to combine.

Never reaching positions of authority or high status the woman potter's knowledge of the work process or comparative wage levels remained minimal, and handicapped even her individual bargaining with her sub-employer. Most women were unskilled and easily replaced given the local surplus of female labour during the period. Only the skilled transferrers could effectively halt the production progress as they did in 1892 and 1900 in pursuit of their wage claims. The sub-employment system was as old as the industry itself and was therefore reinforced by the custom of generations. Divided even amongst themselves by sub-employment, women could only help perpetuate rather than change a system which ensured their subordination.[25]

Few firm generalisations can be made about the potters' wages. Table 4.1 only provides a highly generalised and

Table 4.1: Weekly wages in the pottery industry 1883–1908 (shillings)

Occupation	1883 Wages return Male	Female	1891 Arbitration Male	Female	1908 Poor Law Inquiry Male	Female	Department
Headman					28–40s		Slip house
Assistants					20–5s		
Thrower and	42s		33s		30–50s	9–12s 6d	Potting shops
Assistant	24s	12s			25–7s		
Turner and	33s		25–8s		25–35s		
Assistant		11s				9–13s	
Flat presser	27s	18s	39s		25–35s		
Saucermaker	39s		37s	8–10s	26–38s		
Hollow-ware presser	30s		35–7s		31s		
Handler	30s	6s (girl)	25s	6s	20–4s	10–12s	
Mould-maker	39s		36s				
Head fireman	27–42s				40s		Firing dept
Ovenmen	40s		28s		16–45s	(assistant) 7–10s	
Dipper					35–45s		
Warecleaner		8–12s				10–14s	
Printer	27s		29s		28–30s		Printing
Transferrer		12s		11s		12–14s	
Cutter		4–7s 6d				5s	Decorating
Groundlayer	30s	18s			24s	12–16s	
Painter		18s			25s	11–16s	
Gilder		8s				10–13s	
Burnisher						8–10s	
Head Warehouseman	27s				25s		Warehouse
Assistants	7s (boy)	7s			18–20s	10–13s	
Packers	30s				25s		

Sources: The Return of Wages (1887), C5172, 1883, Return for North Staffordshire, p.236. The 1891 Arbitration Report, The Pottery Gazette, 1 June 1891. The Royal Commission on the Poor Law and Relief of Distress, 1908, Appendix XVI, pp.159–60.

approximate picture. Wages within the same occupation varied widely due to the complex variety of product, plant, ware size, counts and design. Two hollow-ware pressers working in adjacent workshops on one potbank could be receiving weekly wages differing by 10 and sometimes 15 shillings. The detailed survey of Squire and Steel-Maitland of seven pottery firms in 1907 reveals the wide variation in wage levels both between departments in one firm and between different firms and branches of the industry.[26]

In broad terms a potter's wages reflected his or her location within the division of labour, level of skill and status in the potbank. The wage hierarchy reinforced the divisions of the workforce derived from the labour process. In any potbank the fireman, head slipman, some hollow-ware and most sanitary pressers would receive the highest wages of around £2 per week. The second highest paid group consisted of the main crafts such as the throwers, turners, most pressers, mould-makers and printers on 30–8 shillings per week. Various types of skilled male workers could earn between 20 and 29 shillings including placers, groundlayers, slipmen and handlers. Unskilled adult males and some women, notably the skilled transferrers, are recorded as receiving between 10 and 18 shillings. From as low as 3 to 10 shillings was the wage of the majority of unskilled women and juveniles. As befitted her subordinate status, the average weekly wage for all women was only a third of the skilled male potter's rate. Even unskilled men potters could receive double a woman's average wage. The skilled female presser doing virtually the same flat work as a man was paid two-thirds of the male rate.[27]

The earnings of a potter bore little resemblance to the wage levels recorded in official statistics. Earnings, like wages, varied widely and this was due to a number of factors. In an industry where bottlenecks were common, no work meant no pay. A woman tower explained in 1908 how 'There are hours and hours that I am on these works that I never get a penny for . ... many a day we lose two or three hours'. Earnings were also seriously affected by deductions. The Departmental Committee on the Truck Acts in 1906 discovered the industry had a system of wage deductions 'of the longest standing which we have come across'. Besides the sub-employers having to pay assistants out of their own wages, deductions were made by firms for good-from-oven, heating, light, tools, sweeping and eating and washing facilities.[28] Employment was notoriously irregular due to the market conditions and reliance of the industry on coal. The coal stoppage of 1893 on top of the bad trade of 1892 brought the potters to 'the verge of starvation'. The Royal Commission on the Poor Law concluded that in the pottery industry, 'irregularity was the chief evil. Many potters, nominally on good wages, were not really getting a good average wage.'[29]

Consequently the earnings of most potters and especially the unskilled grades were extremely poor. Table 4.1 shows how little wages had moved between 1883 and 1908. The skilled and craft potters were similarly affected. One of their number remarked in 1890 how they had 'suffered more than any skilled industry of late years . . . and in many cases had their prices and conditions of labour so much interfered with that many potters occupy far worse positions than their fathers did'. Female potters' earnings were the lowest of all. Many worked as casual labour or had highly irregular work as assistants with no uniform level of earnings from week to week. One survey revealed that over 53 per cent of the women's earnings it looked at fell between 8 and 12 shillings, while 32 per cent earned only 6–8 shillings. Mary Marland thought the women's wages were 'miserably scandalously low — they were hardly worth the name of wages'. Compared with other women workers, the women potters were amongst the worst paid in the country. Of the fourteen principal women's industries, the women potters were ranked eleventh in terms of pay.[30]

Low wages and earnings were major obstacles to the development of strong trade unions among the potters. Even the National Order of Potters, based upon the flat-pressers, could not grow in numbers in the depressed conditions of the 1890s. William Owen had to admit in 1898 that 'their earnings had sometimes made it difficult to pay even the subscription to a trade society'. Similarly the Women's Trade Union League found the union due to be 'cruelly hard to pay out of such wages'. The WTUL branches could only give a 3 shilling a week out-of-work benefit and could never afford a sick or funeral fund. One presser saw the low earning levels of his craft as the prime cause of the weakness of his union. He wrote in 1891: 'with the poverty prevalent in the cases of the majority of potters, . . . it is clearly manifest the closer the workman has to keep to his labour in the hope of earning a small pittance, the less he thinks of his share of the profits'.[31]

The working conditions of the potters added to their difficulties in maintaining their unions. The ability to attend to union activities or duties was diminished by the long hours of work. Seven in the morning until six in the evening, with one and a half hours for meal breaks, was common. One trade unionist describes the problems facing the potter in individual wage bargaining:

> The articles in the trade are multitudinous, the processes various, the general facts difficult to ascertain and so much time, labour and aptitude are required to prepare the evidence that it is impossible for those working at the bench to do the necessary work of this preparation . . . if a potter holding a situation must give the necessary time to such work, of course he would have to suspend his employment.[32]

Women had the additional burdens of a domestic workload. Eliza Bickerton, a gilder, died of lead poisoning on 21 January 1900. At her inquest it was revealed that she rose on the morning of her death at 5 a.m. to prepare her husband's breakfast. Women potters were forced to do their shopping at night with the local markets remaining open especially for them.[33]

Pottery manufacture was one of the most dangerous of the so-called 'dangerous trades'. The Registrar General's figures for 1980–2 show pottery manufacture had the highest mortality rate of all the dangerous trades. With a mortality rate nearly double the national average the Factory Inspectorate's figures also classify pottery as one of the most dangerous occupations in the country until 1906. As late as 1910 the industry had the highest number of female workers officially suspended from work due to ill-health, of all the dangerous trades.[34]

The two main causes of the potters' poor health were dust and lead. Chest diseases accounted for 58 per cent of the total mortality of potters. Dr Arlidge, during a lifetime's work among the potters, revealed how the silaceous particles within the clay dust damaged lung cells, air tubes, the mucous membranes and rendered breathing organs useless. Potters' rot or phthisis was accompanied by acute pain and suffering. The cases of sufferers and the potters' complaints regarding their suffering were legion and well recorded. One potter spoke for many when he wrote how

> potters die not through accident, but by slow yet certain degrees, not suddenly, but many of them are condemned to a living death and an early grave. There is, metaphorically speaking a skeleton with nearly every 100 crates of ware sent out of the district and the worst tariff that working potters have to contend against is the death tax.[35]

For the potter who worked 'in the lead' the effects could be equally damaging. Lead compounds were key elements in pottery glazes and colours. Aided by the ancient potbanks and their consistently poor sanitation, lead was easily taken in by the potter during work. The effects could be horrific. In its milder forms, gastric disturbance and pathological changes in the blood were the main symptoms. Paralysis, blindness and degeneration of the brain characterised its more acute forms. Even the Home Office statistics indicate the prevalence and extent of lead poisoning among the potters. What they fail to demonstrate is the full impact of the affliction. Studies by Drs Arlidge, Prendergast and Moody, the WTUL, Hanley Labour Church and the SDF all show that the official figures recorded only cases reported to certifying surgeons: thus the Home Office figures were a serious underestimate. Gertrude Tuckwell further demonstrated how the attack rate for women was 2–3 times the male rate. Women's reproductive system and general health were viciously attacked by lead. Of children born to women lead workers 38.4 per cent died in infancy in 1899. Between 1845 and 1905 the Potteries had one of the highest infant mortality rates in Britain.[36]

Pauperism was closely connected with the unhealthy

nature of the pottery trade, and was characteristic of the lives of many potters, especially in old age. Even in 1909 lead poisoning and its effects led the WTUL to speak of the 'great distress where the disease does occur . . . from the extreme poverty of the sufferers.' The entry for a man, 'A.C.', in the Hanley Labour Church survey of 1899 was typical:

> Starting work at 15 after a month he became paralysed. He was twice admitted to the infirmary but nothing could be done for him. . . . He is quite helpless and unable to follow any occupation. Permanently a physical wreck.

In the absence of sick clubs and at a time of high unemployment in the 1890s, the effects of illness were acutely felt. As a Majolica painter explained, when struck by lead poisoning she and her family 'had to manage. The neighbours were very good and helped the children. My husband was not at work and he looked after me. . . . Yes, we got into debt for nourishment.'[37]

To belong to a trade union was immensely difficult when thus afflicted or unemployed due to ill-health. Before many potters could exercise the deliberation and self-denial necessary for trade unionism to flourish they would have needed to enjoy a basic standard of health, surplus energy and leisure which few attained. Catherine Mallet's study of the pottery industry and other dangerous trades revealed their difficulties. As she said to them, 'it would be as reasonable to urge combination upon the patients in a hospital ward'.[38]

Even the highest paid potters had their earning power and regularity of employment disrupted by ill-health. The ovenmen, for example, were highly susceptible to the ravages of heat and dust. Callear, the ovenmen's leader, fought a life-long battle against the conditions which prevented most ovenmen from working beyond 40 years of age. Jabez Booth used the example of Thomas Pickin, a craftsman hollow-ware presser, in his evidence to the Samuel Committee on Compensation for Industrial Diseases. Pickin suffered from phthisis, 'but he could keep on working now and again, as is the case with many potters.

They work part time, and are ill 2 or 3 days, and may then
scramble to work again for 2 or 3 days, it is positively a pain
to see how some of the men suffer.'[39]

As the Webbs pointed out, the potters' unions 'included
but a small percentage of their trades'. In the 1890s the
combined strength of their unions amounted to 6000 out of a
potential membership of 37,000. Though the potters were
weakly organised, in aggregate terms some of the craft
unions were relatively strongly organised within their craft
and especially around disputes. One potter predicted that
the main unions' strength would treble in 1891, if a strike
took place. Indeed in the next year the membership of the
National Order of Potters rose from around 500 to 1400. The
printers and transferrers during their dispute in 1900 had
over half their trade within the society. Yet as the NOP's
rule-book admitted, with regard to the 'history of union
movements in the pottery trade . . . sectional branches have
not developed the necessary power'.[40]

The diversity of potters' unions was a direct outcome of
the group interests apparent within the division of labour.
Each society served a specific craft or occupational group.
The modellers met in 1901 to form a society purely to
protect their own immediate interests. Craftsmen in separate branches of the industry formed their own unions. For
example, the 320 jiggerers in the China trade formed a
society in 1901 entirely separate from jiggerers in other
branches of the industry. It is possible to identify sixteen
main unions within the industry during this period. Closer
analysis reveals two groups within that number. The first
group was composed of the more powerful craft unions
whose membership and funds allowed them a continuous
existence from the middle of the nineteenth century to 1905.
In the second are found a number of societies which
attempted to serve the smaller groups with low membership
levels and uncertain prospects.[41]

The first group consisted of the societies formed by the
hollow-ware pressers, sanitary pressers, flat pressers, ovenmen and the printers and transferrers. These were based on
the major craft groups within the industry. All had built up
lodge and branch networks and a continuous leadership

structure which covered the Potteries: most, notably the ovenmen, had long traditions of organisation dating back to the 1840s. These successfully operated unemployment and strike funds and paid higher benefits than the other unions. Both the hollow-ware pressers and ovenmen had sick benefit and funeral funds. By contrast the second group of unions were based on the smaller, mostly skilled sub-groups within the division of the labour. Their potential memberships were small and their existence highly unstable. The cratemakers, for example, had only a possible maximum membership of 400 in 1890. In the previous decade they had tried to establish a society almost annually.[42]

Outside of these two groups of unions the remainder, the vast majority of unskilled pottery workers, were left almost completely unorganised. Although the National Order of Potters and the hollow-ware pressers accepted certain other skilled clay workers in their societies, around the 1891–2 dispute, they never sought to organise effectively their unskilled sub-employees or assistants. Given the difficulties women potters faced in becoming trade unionists the *Staffordshire Knot* recorded in 1890 that 'thousands of women are destitute of the benefits of organisation'. Clementina Black discovered that as regards employers, 'the women seem powerless in their hands'. All the male unions combined never had more than 400 women members. The mixed printers and transferrers had the highest number with a total of 1035 women members in 1900. The branches of the North Staffordshire WTUL reached a membership of 500 in 1894. Even these figures fluctuated with the WTUL membership reduced to 50 in 1897 and the printers and transferrers down to 305 in 1904. At most the number of unionised women potters amounted to barely 5 per cent of the total female workforce during the period. Gertrude Tuckwell thought in 1897 that, given their numbers, trade unionism would have flourished among the women potters. After her work among them she wrote: 'there is no place in Britain where it ought to be easier to achieve it, for to quote a member of the union, the trade in the potteries is really in the women's hands'. In practice the women potters were partly reflecting the national situation. Between 1896 and

1901, outside of textiles, only 30,000 women were in trade unions.[43]

Amalgamation of the potters' unions was never realised within the period. Full amalgamation did not occur until 1920. In that year the Profiteering Committee rightly stated that 'for many years the operatives concerned in the Pottery Industry were members of a number of small societies or unions, each self-contained and working towards its own ends'. The craft unions even found difficulty in unifying their highly autonomous lodges in each of the Six Towns, or combining the actions of separate trade committees within their ranks. Amalgamation of all the unions or even the major societies was immensely difficult. The call for unity in 1890 by the NOP was itself contradictory, and highlighted the divided allegiances of the potters. It declared, 'potters can rescue their rights and interests if they will stand shoulder to shoulder, and *individually as branches*, make the welfare of each man in the trade the standpoint of action for a united army of potters'. A federation was formed in 1891 but quickly folded when tested by the divisive issues of the 1892 strike. Many potters saw their lack of union unity as probably their central weakness. One writing in 1892 was typical:

> I want to see the day when all the various societies shall sink their petty jealousies and unite together in one grand united society of potters, which could accomplish more good in six months than has been accomplished in the same number of years.[44]

The main cause of the potters' failure to amalgamate was sectional craft interest. Disputes among the craft unions were common. In 1890 for example the throwers, turners and handlers met to discuss the possibility of an association covering the potteries: it ended up in a free fight. Similarly the hollow-ware and flat pressers (although amalgamating in 1899) failed to act in concert on a number of occasions due to fundamental disagreements over their respective rules and practices. Joint action was possible between the main union leaders over the dust and lead regulations of the early 1890s. Even then the dippers opposed the campaign fearing that government regulations would lead to unemployment among their members. Faced with technological change, each society looked to preserve its own craft interests. Throwers and turners fell into dispute over the jolly, a semi-automatic profiling tool used in shaping the clay, while hollow-ware and flat pressers had completely opposing views on its introduction. As a rank-and-file presser said in 1890, it 'was their readiness to quarrel that made unionism so difficult to consolidate and make permanently strong'.[45]

Craft consciousness was strong among the major unions. The transition from a purely domestic craft to factory-based production had been completed without large-scale mechanisation; the necessary innovations were in bodies and glazes. Throughout the nineteenth century the craftsmen's position remained central. Their unions developed around the protections and maintenance of craft privileges and controls as their rule-books show. Each sought to limit entry to their respective craft by apprenticeship. Status, income and benefits were maintained by both the formal and informal sets of controls which the craft potters constructed. The printers and transferrers, for example, stipulated, that apprentices had to serve a five-year apprenticeship and not more than one apprentice to five journeymen was allowed.[46] Craft consciousness was displayed in their union activities. During the 1890 unrest over wage levels 'E.B.' thought 'his duty to himself, his family and his class is how to raise his craft so that it will yield him a living rate of wages'. Another 'maker' thought his craft demanded that he possess 'the greatest skill and have a touch as delicate as that of a lady, yet the strength of a navvy' to produce 'products of pure handicraft'. Such men fought for their privileges consistently. The 1892 strike originated over the question of limitation of apprentices. Every craft union leader at the 1891 wage arbitration based his case on the defence of established craft rights.[47]

The craft leaders were conscious of their status and power within the industry and the community. As befitted their position, it was the craft leaders who represented the workforce in any dealings with government, arbitration boards and the North Staffordshire Trades Council. Owen, Parkes and Edwards of the NOP, printers and transferrers, and ovenmen respectively were three examples of craft union leaders who became local councillors during this period. As elsewhere, the main craft officials sought the public approval of the more important manufacturers and dignitaries. The hollow-ware pressers in 1890 had the mayor of Hanley, J. Wilcox-Edge, to speak at their annual dinner. The printers and transferrers' dinner won local praise for its display of 'fine clothes' and 'sober behaviour'. Given the small size of many firms and the modest capital needed to enter the cheaper branches of the industry, it was quite possible for craftsmen to become pottery manufacturers. George Cadman and Edward Hawley did so with £40 capital saved from wages of 27 shillings and renting their premises for only 25 shillings a week. So prevalent was the practice among the hollow-ware pressers that they made a rule whereby members who entered business and failed could return to the society after three months with no loss of benefit. The difference between the craft potters and their unskilled assistants was socially just as marked. It led Charles Shaw to remark how it was 'impossible to unite classes which differed so widely in sentiment and habit'.[48]

The problems presented by technological change increased the divisions of the craft unions and further alienated the potters outside the societies. Relatively unaffected by technological change since the mid-nineteenth century, the potters found its impact in the 1880s and 1890s all the more traumatic. 'This is the machine age, particularly in the potting trade,' was the trade journal's opinion in 1892. Certain crafts were rendered obsolete like the throwers and turners upon the full-scale adoption of the jolly and the jigger, a hand-held aid used to shape the turning clay. Other crafts had their work and status in the potbank threatened. The hollow-ware pressers did not oppose the new machinery itself but the implicit transformation of their work which jigger and jolly suggested. The hollow-ware pressers' executive admitted that 'we have never attempted — knowing it worse than fruitless — to resist and oppose the introduction and development of unproved methods of working, what we have to watch and struggle for is that the toiler, who manipulates and controls these new methods, should not be the losers by their introduction, either on account of wages or by having labour forced upon him'. Unskilled potters used as substitutes for craftsmen displaced by new technology received the full force of the dispossesseds' anger.[49]

The major conflict arose between male craftsmen and unskilled female potters. The demand for women increased markedly as the new machinery de-skilled certain occupations and opened them up to women as in flat-pressing. Women threatened the jobs and status of craftsmen in four main ways. First, women were paid only a half or sometimes two-thirds of craft rates, and so lowered the rate for that job. Joseph Meir, a saucermaker explained how, in the late 1880s, 'the reduced wages for saucermakers is largely due to the introduction of females at some places, and when men apply for employment they are told they will have to work at the same price as women'. Secondly, upon the introduction of female labour, the nature of related craftsmen's jobs underwent drastic change. A woman flat presser was set to the lighter, smaller ware while the men would be left with the larger more difficult articles. To maintain his wage level the man would therefore have to work harder and faster in order to produce the same amount of work as before. Thirdly, the apprentice system was almost destroyed in places. Women and juveniles working the new machines were called apprentices by employers but this merely signified that they were earning low wages. Pickin of the hollow-ware pressers thought that 'some men are forced to work as lads [apprentices] to get work at all . . . the trade is completely over-run with them'. Finally, when employers observed the cost-cutting advantage of female labour the threat of unemployment loomed for the craftsmen. As the Census shows, male potters decreased in number in the 1890s while the women potters increased by 10.9 per cent.[50]

The changing role of the female potter seriously affected the craftsman's position in the workforce. She threatened his economic security and upset his notion of patriarchy. One flat presser was outraged at the introduction of women to his craft, and protested that it 'is degrading to her position, lowers her in the social and moral scale, and deprives her of paying those attentions she owes to herself and her domestic surroundings, and will ultimately inflict upon herself and society an irretrievable injury'. Women were not acceptable to male potters when working for themselves as opposed to sub-employers. In the words of one presser: 'we believe that for women to be employed at the jollies is wrong; when a woman works at a jolly she occupies an arduous and responsible position which is wholly unfit for a woman'. The craftsmen were also annoyed with the employers who, by employing women, had transferred the gains of cheaper labour out of the sub-employers' pocket into their own. A small group of potters (with the help of Hanley UP) did try to assist the WTUL to organise female workers in the industry between 1893 and 1900. But most male craftsmen instead of allying with the female potter to fight for the introduction of machinery on terms advantageous to them both, attacked her. In so doing they were attacking a symptom but not the cause of their predicament.[51]

Given the divisions created amongst the potters by technological change, it is possible to understand why they failed in their attempts at craft control and their protection of all the potters' interests. As a workforce they were unable to establish uniform price lists or counts and allowed the iniquities of 'good-from-oven' and the allowance system to continue until after the first world war. Industry-wide collective bargaining was never fully attempted; instead the demands of the unions expressed only their sectional interest. The craft unions' autonomy and the independence of their lodges meant that the majority of strikes and disputes were isolated and small-scale. Women and the majority outside the unions were left to bargain individually or in *ad hoc* groups. The major strikes and lock-outs of 1880, 1892 and 1900 served to illustrate the weakness and divisions within the potters' ranks. In December 1891, the ovenmen at Meakins were in dispute over apprentices which led to the general lock-out of May 1892 over an even wider range of issues. Different societies pursued different objectives by varying means, and as a result were defeated. Although 640,000 working days were lost by the 1900 strike the flat pressers, hollow-ware pressers and printers and transferrers only received half their proposed advance despite the greatly improved trade of that year. The labour correspondent of the Board of Trade explained the outcome thus: 'the requirements of the various sections were not uniform nor were they united in their action'.[52]

The divergence of the craft unions' interests critically undermined the effectiveness of the Arbitration and Conciliation Board which their leaders valued so highly. The leaders' intention was for the Board to settle only disputes in individual firms or at most branches of the industry, and not to be a forum for industry-wide collective bargaining. The other workers in the industry were therefore unrepresented. It was assumed that sub-employers would adjust the wages of their employees after any decision by the Board. Given this sectional use of the Board when issues covering the whole industry were put to arbitration in 1879 and 1891 the potters failed to win their demands. Playing one union off against another, the employers prevented the wage rises which the potters sought after the reduction of 1879 and the widespread price-cutting of the period. Sidney Webb pinpointed the potters' weakness, advising them to be stronger and above all more representative if arbitration were ever to be of any use to all potters. The Board dissolved twice, once in 1880 and again in 1891, and remained out of action until 1908. Even Owen, the chief architect of arbitration, realised its failure after 1891 admitting that it had only proved a complicated form of registering their disunity.[53]

Admittedly, other factors did affect the form and strength of the potters' unions. The antipathy of certain sections of local religious opinion to trade unionism also hampered combination amongst the potters. The regional and social isolation of the Potteries retarded the development of the entire industry and the Six Towns.[54] Although it has proved useful in the history of labour and in accounts of individual industries, the concept of a labour aristocracy is of minimal use in the case of the pottery industry. It has been shown how wage levels, earnings, regularity of employment, status and conditions of work varied even within the ranks of the most senior craftsmen. Market forces and skill-displacing technology did not produce the preconditions necessary for the emergence of an identifiable, permanent, aristocratic élite. The hollow-ware pressers, for example, though one of the highest paid groups in the 1880s and 1890s, hardly existed by 1910. No group of pottery workers fulfilled even a majority of the features of the labour aristocrat as designed by Hobsbawm.[55]

This study contends that the potters' organisation and actions between 1880 and 1905 can primarily be explained by reference to the industry in which they worked and by which they lived. Neither the industry nor its workforce can be treated as a unified whole. Both were subject to the horizontal divisions of trade group, firm, potbank and workshop. The workforce was vertically divided by sex, age, skill and status which produced its intricate system of sub-employment. The potters' unions therefore present an equally complex picture, varying in definition, form, policy and action. As befitted their position within the division of labour the craft potters formed the most important societies in terms of numerical and overall strength. The key to the weakness of both the craft unions and of the potters' actions in general lay in their many divisions. The divisions of the potters in the potbanks were reproduced by their unions. When faced by the problems produced by low wages, unhealthy conditions, technological change and an adverse market each sector of the workforce and each union acted largely in terms of defensive self-interest. In particular the craft consciousness of the major unions would not permit a united response and in many ways served only to deepen and intensify the potters' divisions.

The divisions amongst the potters' unions were highlighted by each of the three bouts of major industrial conflict between 1880 and 1905. Lord Hatherton's arbitration in 1879 resulted in a 8½ per cent wage reduction for all grades. The principal aim of the potters in the disputes of 1880, 1890–2 and 1900 was to recoup the so-called 'Lord Hatherton's pennies'. In each dispute, sectional interest undermined the attempts to press a united claim. Varying levels of preoccupation with new technology, female labour, and the apprentice system obscured and frustrated the craft unions' main aim. Each episode of conflict saw attempts by some craft union leaders to federate or amalgamate under pressure from their rank and file; some did try to widen their membership, as in 1890. But each dispute ended in failure for most potters, so that projects like the 1879 and 1892 federations were tainted with failure and therefore short-lived. So marked was the defeat of the unions in 1880 and 1890 that those that survived spent the following decades rebuilding their organisations. What resources remained went towards their own survival. It would be expecting a great deal to suppose that the craft unions could have even begun to tackle the wider problems of low prices and low wages and organising the unorganised potters. That the craft unions survived was due mainly to the enduring importance of the craftsmen in the production process. Despite their inability to strengthen their craft control or even improve their wages any reconstruction of trade unionism in the industry was bound to involve or occur around the craftsmen. The experience of the potters' societies from 1880 to 1905 explains why Warburton could write of them as 'clinging to craft unionism long after it is out of date'.[56]

A decisive feature of the pottery industry during this period was the woman potter. Since that time no one had investigated her role in the history of the industry and helped to create and maintain.[57] She must be central to any analysis of the pottery industry and workforce. It is impossible to understand the actions of any of the potters' societies and above all the craft unions, without reference to the female potter. The evidence for 1880–1905 suggests that the woman's customary subordinate position and lack of power within the work group meant that she experienced the lowest levels of pay and status and produced the weakest forms of combination. Her subordination and impotence were maintained by the craftsmen who employed her. Consequently the woman potter — that is, half the workforce — remained largely outside the ranks of the trade unions. The use of women as cheap substitute labour for craftsmen displaced by new technology served only to deepen the divisions between them. As happened else-where, the craftsmen fought to preserve their craft control and power, but at the expense of the majority of potters. Finally, the female potter is an example of how the experience of the woman worker can broaden our understanding of the history of labour. As the American potters reminded their Staffordshire counterparts in February 1895, using the words of Hosea Biglow:

> Laborin' men and laborin' women,
> Hev one glory and one shame;
> Ev'rything thet's done inhuman
> Injers all of 'em the same.[58]

NOTES

All works cited published in London, unless otherwise stated.

1. This study is based on work done in connection with an MA thesis 'The Women Pottery Workers of Staffordshire and Trade Unionism 1890–1905', submitted in October 1979 to the Centre for the Study of Social History, Warwick University. The study ends in 1905. From 1906–1908 a series of amalgamations began which produced the National Society for Male and Female Pottery Workers. As quoted in C. Shaw, *When I was a Child by an Old Potter* (1903), p.183.

2. The collective name given to the area of North Staffordshire made up of the Six Towns of Tunstall, Burslem, Hanley, Stoke, Fenton and Longton is the Potteries. The term potteries here will be used as the plural form of a pottery (a factory where pottery is manufactured). 'Potbank' is a local alternative to pottery factory. National Order of Potters Rule-book (Hanley, 1901), Rules of the Executive Council. H. Barrett-Greene, TUC (Hanley, 1905). *The Programme and The History of the Staffordshire Potteries* (Longton, 1905), p.22.

3. E. Surrey-Dane, 'The Economic History of the Staffordshire Pottery Industry to 1850', MA thesis Sheffield, 1950. H.A. Moisley, 'The Potteries Coalfield, a Regional Analysis', MSc thesis, Leeds, 1950, pp.91–5. *Report of the Lady Inspector of Factories* (1902), p.239. *The Pottery Gazette*, 2 November 1896, p.907. *First Census of Production* (1907), p.750.

4. Memo to the Home Secretary, *Justice*, 30 April 1898. *Census for the County of Stafford*, Table of Occupations of Males and Females Aged Ten Years and Upwards, 1881, 1891, 1901 and 1911. According to p.65 of the 1911 Occupation Table only 14,088 workers were employed in and around mines and 1813 in iron and steel. *The Pottery Gazette*, 1 May 1896, p.359; *The Daily News*, 2 January 1904; *Christian Commonwealth*, 7 January 1904; A.H. Morgan, 'Regional Consciousness in the North Staffordshire Potteries', *Geography*, (March 1942), pp.94–100. S. Baring-Gould, *The Frobishers* (1901), p.78.

5. B. Whitelegge, *Report of the Departmental Committee to Inquire into the Dangers Attendant on the use of Lead in the Manufacture of Earthenware and China* (1910), Cd.5219, 5278, 5385, vol.II, pp.5–6. *Kelley's Trades Directory of Staffordshire*, 1892 and 1900. *The Potteries, Newcastle and District Directory* (Hanley, 1907), pp.8137–914. The Minton Manuscripts, item 643, University College of North Wales, Bangor. F. Parkin, *Autobiography of a Trade Unionist* (MS, n.d. [1920?]), p.xvi, Horace Barks Library Hanley, *The Pottery Gazette*, 2 November 1896, p.907.

6. *The Staffordshire Advertiser*, 11 January 1890, p.4. The local correspondent of *The Pottery Gazette*, 1 September 1892. 'Sweating in the Potting Trade', *The Workman's Times*, 10 October 1890.

7. *The Profiteering Committee*, p.5. Williams, *op. cit.*, p.308. Article on 'The Printers and Transferrers', *The Pottery Gazette*, 1 March 1892. See, the monthly reports on the pottery industry, in *The Labour Gazette, Journal of the Labour Dept. of the Board of Trade 1894–1905*.

8. Mr Entwistle, *Hansard*, 1927, p.851. *The Staffordshire Advertiser*, 28 April 1900. *The Christian Commonwealth*, 7 January 1904.

9. Moisley, *op. cit.*, p.131. Drawing Red, *The Workman's Times*, 27 February 1892. *The Pottery Gazette*, 1 August 1890, 1 May 1893 and 1 June 1899. *Christian Commonwealth*, 7 January 1904. *Sentinel*, 25 January 1901.

10. WTUL, *Annual Reports* 1895, p.5; 1898, p.13; *Quarterly Review*, October 1894, p.6. *The Pottery Gazette*, 12 April 1900, p.419.

11. W. Burton, *Departmental Lead Committee*, q. 17,077. J. Foster Fraeser, *Life's Contrast* (1908) pp.141–55. *The Pottery Gazette*, 1 July 1892, 1 January 1893. *Staffordshire Advertiser*, 11 January 1890. Interview with E. Grocott, 3 December 1979. *Potter's Examiner*, 10 May 1879.

12. Morgan, *op. cit.*, pp.99–100. B.R. Williams, 'The Pottery Industry', in D. Burn (ed.), *The Structure of British Industry*, vol.2, 1958; Report of the Chamber of Commerce, *The Pottery Gazette*, 1 February 1894.

13. Letter from a 'potter' to *The Workman's Times*, 3 and 10 October 1890. WTUL, *Annual Report*, 1892, p.5. Admittedly Arthur Brownfield was chairman of Hanley ILP. Letter from a 'manufacturer' to *The Pottery Gazette*, 1 July and 1 November 1890. F.R. Benham, *Royal Commission on Trade Disputes and Trade Combinations* (1906), LVI, col. 2826, qs. 4979 and 5012.

14. The local correspondent of *The Pottery Gazette*, 1 August 1890, 1 June 1892, 1 September and 1 November 1898. Brownfield, *op. cit.*, p.9. *The Departmental Lead Committee*, q.14,995.

15. William Burton, *Departmental Committee on the Truck Acts* (1906), Cd.142, p.42 and q.17,077.

16. For accounts of the industrial and labour process of pottery manufacture, see *The Departmental Lead Committee*, vol.1, (1910), pp.44–91; and Barrett-Greene, *op. cit.*, pp.67–71.

17. A Hollins, *Improperly Pugged Clay*, National Council of the Pottery Industry (Hanley, 1924). F. Celoria, 'Ceramic Machinery of the 19th Century', *Staffordshire Archaeology*, no.2 (1973), p.46. *Report of the Lady Inspector of Factories*, (1903), p.22.
18. Clara Collet, *Report on the Employment of Women in the Staffordshire Potteries to the Royal Commission on Labour* (1893), c.6893, p.61. 'Death in the Workshop', *The Daily Chronicle*, 14 November 1892, p.7. The term craftsman will be used to denote a skilled worker who achieved commonly accepted craftsman status only after having served a set term of apprenticeship at his craft.
19. W. Burton, *Minutes and the Inquiry into Draft Regulations for the Manufacture and Decoration of Pottery before Judge Ruegg* (1912), 27 November, q.954. Thomas Edwards, agent of the United Ovenmen, Evidence taken before Group C of the *Royal Commission on Labour*, vol.III (1903), c.6894, p.81. *The Departmental Lead Committee* (1910), vol.I, pp.61, 71.
20. Collet, *op. cit.*, p.62.
21. W. Redgrave, *Departmental Committee on the Truck Acts* (1904), pp.714, 778. *Report of the Lady Inspector of Factories* (1904), p.261. Mrs F., *The Departmental Lead Committee*, (1910), vol.II, q.11,626.
22. Collet, *op. cit.*, p.63. *Census for the County of Stafford*, Table of Occupations (1901), pp.69, 79. 'The Morals of the Potteries', *The Spectator*, 11 January 1904. *Report of the Lady Inspector* (1904), p.261.
23. W. Burton, *Departmental Committee on the Truck Act* (1904), p.42. Collet, *op. cit.*, pp.61, 63. *Report of the Inter-Departmental Committee on Physical Deterioration* (1903), appendix V, p.127. A. Anderson, *The Departmental Lead Committee* (1910), qq. 10830–40. *Rulebook of the Amalgamated Society of Hollow-Ware Pressers* (Burslem, 1890), p.10. *Rulebook of the China and Earthenware Gilders* (Stoke-on-Trent, 1890), section 2, The Webb Trade Union Collection, Section C, vol.72, Coll. E.D. Library of the London School of Economics and Political Science.
24. Webb Trade Union Collection, vol.CIX, 'An Appeal to Hollow-Ware Pressers (n.d. [1891]), and vol.XLIV, item 3, p.310. Collet, *op. cit.*, witness 393. Ms Shuter, *The Departmental Lead Committee*, qq. 10784 and 86997. *The Staffordshire Sentinel*, The Barlow Case, 4 and 9 December 1895. Sylvia Pankhurst, as quoted in R. Pankhurst, *Sylvia Pankhurst Artist and Crusader* (1979), p.80. See also E. Hobsbawm, *Labouring Men* (1964), pp.297–8.
25. In 1901 69.23 per cent of female pottery workers were single and yet had their wages determined by the customary wage rates of the family systems. *Census for the County of Stafford* (1901), pp.69, 79. Collet, *op. cit.*, pp.61–2. N. Parkes, *Daily News*, 9 January 1904.
26. Returns of wages published 1830–86 (1887), c.5172, p.236. 1891 Arbitration Report, as printed in *Pottery Gazette*, 1 June 1891. *The Earnings and Hours Inquiry, Report of the Board of Trade* (1906), VII, pp.102·9. *Royal Commission on the Poor Law and Relief of*

Distress, 1908, Appendix XVI, *Report of Steel-Maitland and Squire on the Pottery's Wages*, pp.159(105)–160(106).

27. 1891 *Arbitration Report, loc. cit.* H. Owen, *The Staffordshire Potter* (1901), pp.320–33. *The Royal Commission on Labour*, Appendix III, *Summary upon the Employment of Women*, p.477.

28. *Departmental Committee on the Truck Acts*, pp.43, 52, 650, 685. By 'good-from-oven' any spoilt ware was held to be the responsibility of the workers involved in its production. The wages of those workers were docked accordingly even though the fault may not have been caused by them.

29. W. Owen, speech at TUC 1893, *TUC Annual Report*, (1893), p.84. *Royal Commission on the Poor Law, loc. cit.*, p.158.

30. *The Workman's Times*, 31 October 1890. 'Report on the Potteries'. Collet, *op. cit.*, p.63. 1891 *Arbitration Report*, as printed in the *Pottery Gazette*, 1 June 1891. *Report of the Lady Inspector of Factories* (1911), p.145.

31. W. Owen, *The Staffordshire Sentinel*, 30 November 1895, Meeting of the NOP. *Women's Trade Union League Quarterly Review*, January 1904, pp.7–9. 'Presser', *The Workman's Times*, 21 August 1891.

32. Letter of George Ingleby, Secretary of the Printers and Transferrers, to the Board of Arbitration, 24 October 1890, as printed in the *Pottery Gazette*, 1 November 1890.

33. Baring-Gould, *op. cit.*, p.80. *The Pottery Gazette*, 1 March 1900, p.303.

34. Royal Commission on Labour (1893), vol.III, para. 477. Report of Dr Tatham in the *Report of the Registrar General*, 1895. Dr Arlidge, *The Conditions of Labour in the Potteries, The Committee of Inquiry* (1893), p.4. *Report of the Chief Inspector of Factories and Workshops*, 1903, p.215; 1910, p.227. Miss Vines, *The Departmental Lead Committee*, qs.3994–5. Dr Prendergast, *The Labour Leader*, 25 June 1898.

35. *The Times*, 8 October 1898, p.8. P.T. Arlidge, *Pottery Manufacture in its Sanitary Aspects* (Hanley, 1892). *The Departmental Lead Committee*, vol.7, pp.68, 70, 75, 79, 83, 89, 91, 97. 'Sweating in the Pottery Trade', *The Workman's Times*, 10 October 1890.

36. *Lead Compounds in Pottery. Report to the Secretary of State for the Home Department by Professor T.F. Thorpe and T. Oliver*, p.15. T. Oliver (ed.), *The Dangerous Trades* (1902), p.307. B. Wilson, 'Our Industrial Victims', *Young Oxford*, vol.II, no.14, November 1900, pp.54–55. Sir Charles Dilke, *Hansard*, 13 July, col. 1479. *Royal Commission on the Poor Law*, vol.XVI, pp.151–69. G. Tuckwell, Minority Memorandum, *The Departmental Lead Committee*, p.129. G. Newman, *Infant Mortality: A Social Problem* (1906), pp.21–3, 108–9.

37. *Royal Commission on the Poor Law, loc. cit.*, p.162. WTUL, *Annual Report* (1909), p.20. Wilson, *op. cit.*, vol.II, no.16, January 1901, pp.126–31. *The Departmental Lead Committee*, vol.I, qq. 11789–92.

38. B. Webb, 'The Difficulties of Organising Women', *Webb Trade Union*

Collection, vol.XLVII, pp.15–25. C. Mallet, *Dangerous Trades for Women* (1893), p.19.

39. W. Calleas, Judge Ruegg Inquiry, qq.534–44. J. Booth, *The Departmental Committee on Compensation for Industrial Diseases*, q.9384, 1906, col.3496.

40. S. and B. Webb, *The History of Trade Unionism, 1666–1920* (1920), p.438. 'Fairplay', *The Potteries Free Press and Staffordshire Knot*, 26 September 1891. Webb Trade Union Collection, vol. XLIV, p.347. *Report on Trade Unions*, 1901, pp.88–9. Introduction to the *National Order of Potters Rule Book*, (Burslem, 1891).

41. *The Staffordshire Sentinel*, 21 February, 19 July 1901. A craft union is defined here as a society which only admitted craftsmen who had served, or were serving, the appropriate apprenticeship for their craft and which thereby protected that craft's status.

42. Hanley Lodge Book of the Hollow-Ware Pressers Union, 1855–1884, the Burslem Lodge Book, 1864–1890, and Webb Trade Union Collection, vol.XLIV, pp.318–24, 342–4, 396. (The slipmakers were members of the Hollow-Ware Pressers' Union.) *The Potteries Free Press and Staffordshire Knot*, 20 September 1890.

43. *The Staffordshire Knot*, 3 January 1891, 22 November 1890. *Report on Trade Unions* (1902), p.7; 1905–7, p.8. WTUL, *Quarterly Review*, 3 May 1897.

44. National Executive Committee of the National Amalgamated Society of Pottery Workers, Amalgamation Sub-Committee, 2 January 1920. CATU House, Hanley. *The Profiteering Acts 1919–20 Central Committee Report on Pottery*, 1921, p.8. Webb Trade Union Collection, vol.XLIV, pp.278, 318. *The Staffordshire Advertiser*, 11 October, p.5. A. Brownfield, *The Lock-Out, A Potters Guild* (Hanley, 1892). 'Fairplay', *The Workman's Times*, 27 February 1892, 29 August 1890. *The Staffordshire Knot*, 26 September 1891.

45. Letter from 'a potter' *The Workman's Times*, 30 January 1891, 3 and 10 October 1890. *The Staffordshire Knot*, 26 September 1891. Webb Trade Union Collection, *loc. cit.*, p.332.

46. H. Moisley, *op. cit.*, p.91. *Royal Commission on Labour, Group C, Answers to Schedules of Questions* (1892), c.6795, pp.70–1, 153–4, 230, 322; *Group C, Rules of Associations of Employers and Employed*, pp.236–8. Webb Trade Union Collection, section C, vol.72: *Rules of the China and Earthenware Gilders Union* (Stoke, 1890) section 2, vol.73; *Rules of the Operative Cratemakers Society* (Hanley, 1890); *Rules of the United Firemen and Kilnmen's Labour Protection Association* (Hanley, 1890); *Rules of the Amalgamated Society of Hollow-Ware Pressers* (Burslem, 1890); *Rules of the National Order of Potters* (Burslem, 1891).

47. Letter from E.B., *The Workman's Times*, 28 October 1890; 'Sweating in the Pottery Trade', 16 October 1890; letter from 'an old sufferer' 28 November 1890. Verbatim report of the Hinckes Arbitration, *The Pottery Gazette*, 1 June 1891.

48. See the *Royal Commission on Labour, loc. cit. Pottery Regulations*

Inquiry, loc. cit. Report of TUC (1905), p.11. *The Staffordshire Knot*, 1 February 1890. *The Potteries Examiner*, 30 October 1875. *The Workman's Times*, 10 October 1890. *The Staffordshire Advertiser*, 28 April 1900. Rules of the Hollow-Ware Pressers, Rule 24, *loc. cit.* pp.193–4.

49. A. Lamb, 'The Press and Labour's Response to Pottery Making Machinery', *Journal of Ceramic History*, no.9, 1977. Local Correspondent, *The Pottery Gazette*, 1 September 1892. Hand Bill of Hollow-Ware Pressers, Webb Trade Union Collection, vol.CIX.

50. Hinckes Arbitration Report, *loc. cit.* Webb Trade Union Collection, vol.XLIV, pp.332, 353–4. Minton Manuscripts, vol.I, item 591, registration of apprentices. *Census for the County of Stafford*, Table of Occupations, 1871–1911.

51. Flat Pressers Appeal, *The Potteries Examiner*, 9 November 1879. Webb Trade Union Collection, *loc. cit.*, pp.301–2; vol.CIX, p.15. Letter of W. Owen, in Brownfield, *op. cit.*, p.10. Even the campaign for factory and health legislation was used by the craft unions to try to exclude women from jobs considered to be solely for men. *The Staffordshire Knot*, 25 October 1890.

52. Pottery Industry Wages Structure (November 1946), p.15. sec. 76. *The Workman's Times*, 6 March 1891. Webb Trade Union Collection, vol.XLIV, pp.175, 276. Local Correspondent, *The Pottery Gazette*, 2 May 1892. *Report on Strikes and Lock-Outs*, 1900, pp.68–9.

53. Webb Trade Union Collection, *loc. cit.*, 175ff. *Rules of the North Staffordshire Arbitration Board* (Burslem, 1885). W. Owen, *The Pottery Gazette*, 1 July, 1 August 1891.

54. Shaw, *op. cit.*, p.192. A. Bennett, *These Twain* (1916), p.192. S.H. Beaver, 'The Potteries: A Study in the Evolution of a Cultural Landscape', *The Institute of British Geographers, Transactions and Papers*, no.34, June 1964, pp.1–31.

55. Hobsbawm, *op. cit.*, pp.272–315. T. Matsumura, 'The Flint Glass-Makers in the Classic Age of the Labour Aristocracy, 1850–1880', PhD thesis, University of Warwick, 1976. Dr C.L. Sutherland and Dr S. Bryson, *Report on the Incidence of Silicosis in the Pottery Industry*, (1926), p.15.

56. Webb Trade Union Collection, *loc. cit.*, p.260. NSPW 1919 wage claim questionnaire. W.H. Warburton, *The History of Trade Union Organisation in the North Staffordshire Potteries* (1931) p.15.

57. S. Webb in his research sensed the importance of the craft unions within the pottery industry and their relations with the female potters through no mention of his findings appears in *The History of Trade Unionism* (1920). Appendix II, if H. Owen in *The Staffordshire Potter* (1901) was aware of the importance of female labour. Since then two outline histories of the potters' trade unions covering 150 years have appeared. However, W.H. Warburton, *op. cit.*, p.213, and F. Burchill and K. Ross, *A History of the Potters' Union* (Hanley, 1877), p.148, make only passing reference to what they term 'the

problem of female labour'.
58. E.H. Phelps Brown, *The Growth of British Industrial Relations* (1959), p.235. S. Rowbotham, *Hidden from History* (1973), ch.12. M.A. Hamilton, *Women at Work* (1941), p.162. J. Druker, 'Women's History and Trade Union Records', *Bulletin of the Society for the Study of Labour History*, no.36, Spring 1978, pp.28–35. As quoted in Owen, *op. cit.*, p.242.

5. Wood, Iron and Steel: Technology, Labour and Trade Union Organisation in the Shipbuilding Industry, 1840–1914.

Keith McClelland and Alastair Reid

The shipbuilding industry is perhaps best known among social and labour historians as the site of the most privileged and craft-exclusive skilled workers in nineteenth-century Britain. Shipyard trades were prominent among the 'super-aristocrats' with very high wages in Eric Hobsbawm's account of the labour aristocracy, and even in Henry Pelling's critical response, the boilermakers appeared as a 'highly aristocratic' group.[1] Robert Knight, who was the general secretary of the Boilermakers' Society from 1871 to 1899, is frequently taken as the standard example of moderate trade unionism in the period, and his remarks before the Royal Commission on Labour, in the early 1890s, have become almost legendary. One of the most frequently quoted is the following comment on 'craftsmen' and 'labourers': 'the plater is the mechanic and as a matter of course the helper ought to be subservient and do as the mechanic tells him'.[2] We hope to show that such statements, taken out of context as they usually are, can substantially distort the complexity of the relations between workers in the shipbuilding industry. After all, the platers were only one section of a much larger trade union which was itself only one out of over twenty significant unions operating in the industry.

The first part of our essay outlines the main aspects of the economic development of shipbuilding in the period stressing that, while it was expanding rapidly, this sector was also subject to very intense cyclical fluctuations in output and employment. As a result skilled workers' average earnings were considerably lower than has usually been assumed. In the second part we look more closely at the transition from wooden to iron hulls between 1840 and 1880 and the consequent displacement of shipwrights by boilermakers as the core of the labour force. Within this context it becomes clear that skilled metal workers in the shipyards were neither straightforwardly collaborating with their employers nor exercising untrammelled authority over their less skilled assistants. Finally, in the third section, we analyse the later nineteenth century, when steel replaced iron as the main hull building material and foreign competition began to increase. By this period the main unions of the skilled had managed to construct institutional defences strong enough to resist what increase in employers' pressure there was, and as a result there was no extensive mechanisation in the industry. Throughout the period covered by this essay, shipbuilding remained dependent on a wide range of manual skills and aptitudes and the distinction between occupations was never technically straightforward. Relations between the 'skilled' and the 'unskilled' were complex and are best seen as part of a wider phenomenon, occurring at all levels of the workforce, involving the defence of hard-won territory against both the employers and neighbouring occupational groups.

I

The shipbuilding industry was one of the major capital goods sectors which became so prominent in the British economy between the middle of the nineteenth century and the first world war. Like the railways, it had a particular function as an assembly industry consuming the raw materials and components produced by the coal, iron, steel and engineering industries, and indeed after 1880 it seems to have taken over from railway construction as the central pivot of the country's heavy industries. This importance in the structure of the domestic economy was strongly reinforced by the increasing contribution to the balance of payments made by invisible earnings coming from Britain's domination of world trade and by the prominence of warships in the maintenance of imperial security. However, despite its

obvious importance, the definition of the industry's boundaries was rather unclear, precisely because it was so
interdependent with the other capital goods sectors geographically concentrated in the same regions. This overlap
was especially marked in the case of the boundary between
shipbuilding and engineering which gave the construction of
marine engines and other fittings a very ambiguous status,
and led to the contemporary statistical classification 'shipbuilding and engineering' which so bedevils the sources. In
this chapter we have taken the shipbuilding industry to
include commercial shipyards (building both merchant and
naval vessels) and marine engineering and boilermaking,
whether conducted within shipyards, in close association
with shipyards, or in entirely independent enterprises. Our
definition therefore excludes the government-controlled
dockyards and specialist ship-repairing firms in both of
which working conditions were quite different from those in
the commercial building firms, and which tended to be
regionally distinct, remaining focused in the traditional
wooden shipbuilding centres in the south and south-east of
England.

As this distinction indicates, the technical transformations
in the product which resulted in the replacement of wooden
sailing ships with steam and iron/steel ships had a profound
impact on the organisation of the shipbuilding industry. The
adoption of steam power was largely due to technical
progress in the engines available and the shortening of sea
passages by canal construction; meanwhile independent
dynamics, including the movements of raw materials prices
and the assessment of the design and construction advantages, produced a shift from wood to iron and then steel as
the main hull building material.[3] These parallel and inter-
connected processes, which began in the 1820s and were
largely completed by 1890, had a significant impact on both
the scale of shipyards and their geographical location. While
there had been some very large wooden yards, especially on
the Thames, the average size of the firms scattered around
the country's coasts and rivers was nearer to twenty men
with very little fixed capital. In marked contrast even the
earliest iron shipbuilding yards employed from 500 to 1000

men, and had capital of from £5000 to £25,000, while by the late nineteenth-century shipyards were among the largest industrial enterprises in the country with twenty employing over 2000 men each.[4]

At the same time there was no very strong drive towards the centralisation of capital and, with the exception of some significant mergers in the 1900s, ownership remained widely scattered among over 100 predominantly family firms. Similarly, despite the large scale of production, and the huge amounts of capital committed to buildings, land and raw materials stocks, there was no powerful tendency towards mechanisation in the metal shipbuilding industry and production methods remained clearly dependent on skilled manual labour. The wood-working shipwrights were not transformed into, or replaced by, a semi-skilled factory-type workforce, but remained as a highly skilled, if less central, section to which a large range of other skilled trades were added.

The other major change which *did* occur during the transition from wood to metal was a marked geographical concentration of the industry on the Clyde, the rivers of the north-east coast and several scattered towns in the north-west of England and in the north-east of Ireland. This can be explained largely in terms of the advantages gained from locations close to other capital goods sectors which could supply the necessary raw materials and components at competitive prices and which were already creating expanding regional pools of the necessary skilled labour.[5] Of course, many of the prominent metal shipbuilding firms had long histories going back to the period of wooden hulls, and indeed many of the necessary technical experiments and innovations had been carried out within already existing yards on the Clyde and the north-east coast. However, the regional shift was a major one, resulting in the rapid decline of the previously important Thames shipbuilding industry, and the emergence of firms on the northern rivers which no longer depended solely on local demand but increasingly supplied the whole world.

The wooden shipyards of early nineteenth-century Britain had in fact been rather vulnerable to foreign competition,

especially from firms in the United States which had access
to better and cheaper supplies of wood, and from firms in
France which tended to produce more advanced designs.
But because of the predominance of British iron and
engineering until the 1870s the shift away from wood and
towards steam and iron gave British yards the opportunity to
seize a virtual monopoly of the world market in ships which
was only gradually and partially eroded by the USA and
Germany after 1900. This position was also based on the
very large domestic demand for ships arising from the
domination of world trade by British shipping for, even after
British manufacturers had begun to face foreign competitors, British shippers remained the world's main cargo
carriers. As a result British trade continued to expand
rapidly even after British industry had begun to falter:
between 1870 and 1914 the volume and value of trade with
foreign nations doubled, while improvements in ships and
their handling more than trebled the annual carrying
capacity of the British merchant fleet. On the basis of the
large domestic market thus available, British shipbuilding
yards were able to establish economies in production which
were denied to their overseas rivals. They had access to
better and cheaper steel as a result of the large proportion of
domestic steel production which took the form of ship
plates, and they had access to cheaper and more highly
developed components as a result of the high degree of
specialisation amongst suppliers, especially in the case of
marine engines.

The size of the shipbuilding industry thus became a
self-confirming advantage which it was impossible for
foreign yards to break through until the disruption of the
world market during and after the first world war, and it
produced a significant degree of specialisation among
shipbuilders both between and within regions.[6] As a result
of its long standing connections with the coal trade and its
proximity to Europe, the north-east coast developed a
tendency to specialise in general purpose tramp cargo
vessels, while the Belfast and Glasgow yards, with their
convenient location near the North Atlantic routes, were
notable for their output of high-class passenger liners.

However, this contrast was by no means an absolute one, and most shipbuilding rivers could claim a wide range of types of vessel and a number of specialist producers, most notably the largest yards which built enormous liners and warships.

The cumulative effect of these various advantages was that, even in the cases where foreign shipping lines could challenge the British monopoly, it was frequently on the basis of vessels built in Britain, and even purchased secondhand because of their reputation for high quality. While by no means immune from cyclical fluctuations or from alterations in the rate of growth, the general tendency of the industry in the period under consideration was therefore one of secular expansion of output and numbers employed up to their record peak in 1913, and even in cases where mechanisation or the reorganisation of the division of labour produced savings on labour costs in specific processes there was little threat of a net displacement of skilled labour.

This background of long-term expansion of the industry goes a long way towards explaining the remarkable strength of shipbuilding workers' trade unions in the period, especially that of the boilermakers' society which had managed to organise 95 per cent of the eligible skilled and the semi-skilled workers by the early 1890s.[7] This apparent expansion and strength might lead to the misleading conclusion that shipbuilding workers were a straightforward example of a secure labour aristocracy, but this would be to overlook their exposure to severe alternations between periods of very intense work and periods of extended high unemployment. For, despite the partially counter-cyclical impact of state contracts for naval vessels, the industry experienced perhaps the most intense periodic fluctuations in production of any sector in Britain at the time.[8] In the first place, like all the capital goods sectors, it was exposed to the full impact of the world trade cycle as a result of the high elasticity of demand for its products, in marked contrast to such consumer goods sectors as food and clothing in which demand was more stable from one year to the next. But, in addition, it also suffered adversely from the very long construction period of ships and from the customer-specificity of its tailor-made products. This latter feature made production for stockpiling during depressions an unattractive option, while the long period of construction meant that there was frequently an over-supply of ships at exactly the time when the boom was most vulnerable.[9] As a result of the severity of fluctuations in output, the employers preferred labour-intensive techniques. Even as the product became larger and more technically sophisticated, and yard sizes expanded rapidly with the move away from wood, the commitment of resources to tools and machinery was restricted in an attempt to limit overheads and place the burden of the recurrent intense depressions squarely on the backs of the labour force.[10]

The sharp fluctuations in employment which followed on this strategy led to similarly intense fluctuations in the levels of individual earnings, even in the cases of the most skilled and the best paid. The 1906 Board of Trade figures[11] used by Hobsbawm and others are therefore unreliable for ship-building, as they were based on returns of earnings for only one week in a very prosperous phase of the cycle. Even assuming the accuracy of the statistics, they did not represent income actually available for consumption as the skilled men were working very hard to pay off debts incurred during the last spell of unemployment and, if they were provident, to save up for the next one.[12] Furthermore, Sylvia Price's recent work on the wage books of the Clyde shipyards highlights the inaccuracy of the Board of Trade figures due to their exclusion of those men who had failed to complete a full 54-hour week. The difficulty of organising an even flow of work in the yards combined with the skilled men's customs of taking time off work meant that it was very common to work less than the standard hours. More detailed statistics for Clydeside riveters in 1906 indicate that in contrast to the Board of Trade average of 55 shillings per week, most men in this occupation were taking home around 30 shillings, and almost 30 per cent were earning less than this.[13]

Fluctuations in employment and earnings led inevitably to marked fluctuations in the funds, membership and bargain-ing strength of even the best organised trade unions. This

insecurity in the economic position and power of the
industry's skilled workers was manifest both in the movements of the standard district wage rates which were the
outcome of protracted formal collective bargaining, and in
formal and informal conflicts at yard, district and national
level over questions concerning the organisation of work.
This was evident, for example, in the case of the boilermakers' control over the central question of the number of
apprentices to be allowed into the industry, which was more
or less equivalent to the degree to which they could regulate
the size of their occupational labour market. As a result of
the employers' desire to expand that market by increasing
the number of apprentices there was a persistent struggle
around this issue beginning during the rapid expansion of
the industry in the 1870s and continuing up to (and beyond)
the outbreak of the first world war, with the initiative
passing from side to side with the ups and downs of the trade
cycle and of union strength.[14] Thus while the long-term
expansion of the labour force is a part of the explanation of
the strength of trade unionism in the industry, an equally
important part must be the intensity of cyclical fluctuations
and the consequent need on the part of the skilled workers
for strong institutions which could accumulate large benefit
and strike funds to guarantee their livelihood and to protect
their positions in the division of labour. This basic ambiguity
between security, strength and restraint on the one hand,
and insecurity and militancy on the other will be a recurrent
theme in our more detailed discussion of skilled shipyard
workers' positions in the division of labour.

II

As we have indicated, the British shipbuilding industry
underwent major growth and structural change between the
1840s and the 1880s. The consequences of this for labour
were radical. For what was entailed was the restructuring of
the division of labour in the industry, although the way in
which this happened was in marked contrast to the experience of many of the other dominant industries of nineteenth-

century Britain. The most revolutionary path to the reorganisation of the division of labour was via the introduction of machinery, as occurred in cotton. Another path, to be seen in engineering, was the breaking-down of a pre-existing division of labour into more detailed processes but without the full-scale introduction of machine methods of production.[15] In shipbuilding, change followed neither of these routes. Here what was critical was the creation of new trades and skills and the incorporation of pre-existing ones into a division of labour organised on a new basis. The new trades, notably those of the 'black squad' of iron workers, did not emerge from the breaking-down of the skills of the shipwrights — the major trade of wooden shipbuilding — but were introduced from outside the industry.[16] However, they were not the product, in any simple sense, of technical change. Even though the use of iron necessitated changes in the character of assembly the introduction of wholly new trades was not an absolutely inevitable concomitant. That this happened was due rather to both the strategies adopted by employers and also to the conflicts that developed between key groups of workers in the industry.

The methods and organisation of work in wooden shipbuilding in the early nineteenth century were of a very traditional kind, for in many respects the techniques of building had changed little over the previous couple of hundred years. The main groups of construction workers were the shipwrights, who made the whole carcass of the ship; the joiners, whose main task was the lighter woodwork, especially that involved in cabin-fitting; the caulkers, whose job it was to make the vessel watertight; and the black- or ship-smiths who made all the iron parts. Beyond these the main trades were those ancillary ones, like sawyers who prepared some of the wood, or fitting and finishing ones, like sailmakers and riggers.

Though the techniques and trades engaged in shipbuilding were traditional ones they were certainly not unassailable: block- and mast-making, for example, were becoming separately defined occupations where they had once been part of the shipwrights' work. But there were no technical changes which fundamentally attacked the division of labour and the position of the different groups of workers within it. As a result manual skills, experience and strength were at a premium, not least because of the expertise and judgement required in the work itself. Naturally, there was some degree of planning of the ultimate form of the vessel, but the actual execution of a job needed fine attention to the quality of the wood, its readiness for use, and a mastery of the material in order to achieve the precise shapes and curvatures required, and such matters could not be wholly determined before building got under way.[17]

The exercise of the skills of the shipwrights and others in the yards was embedded in the particular form of relations between masters and men, a relation characterised by the relative absence of direct, day-to-day, control over assembly processes by the masters. That this was so was, in part, due to the persistence of the gang system among the men, a form of organisation that seems to have been particularly strong in London. Working and negotiating as a group with the masters the gang contracted to build a vessel for a given price and took responsibility for the work as they also undertook to hire and pay, out of the contract price, any additional labour that became necessary. As such the shipwrights enjoyed a good deal of autonomy in their work.[18]

But that autonomy was limited, primarily by economic compulsions. There was no technical reason why the shipwrights and others could not build a vessel on their own account without any reorganisation of the existing assembly process. Indeed, at some moments they did precisely that, building either on spec. or for a definite buyer, as happened during some strikes in London, or in times of depression as in Sunderland in the 1830s.[19] But they were subordinated to the masters by two factors: first, their non-possession of capital and means of production, although the leading trades did own their own tools; and second, and concomitantly, their dependence upon the wage — although the form of the wage was not a straightforward piece- or time-wage but, rather, one that was entwined with the price of the commodity. The masters were primarily defined, of course, by their ability to raise and supply capital, both in the sense of finance and also in investment in yards, some of which were of considerable size and relatively expensive. But, as yet, the employers had not taken full control over the immediate processes of production.

In the first half of the nineteenth century the shipwrights were coming under increasing pressure from a variety of causes: the industry was more subject to competition, especially from American builders, at the same time as population growth and post-1815 economic dislocations were hitting real wages and standards of living. The shipwrights' response, in a situation also defined by wide-spread political radicalism and the growth of the ideas and practices of trades' organisation, was to form their own trade unions and other institutions of defence. Central to the unions were the attempts to prevent encroachments by employers on the complex rules and regulations governing the trade in matters of apprenticeship, customary wage levels, allowance payments and to prevent the incursion of 'illegal' (non-union) men. Naturally, the ability to sustain organisation and the degree of success which they had varied from port to port according to prevailing conditions: in London, for example, the Thames Shipwrights' Provident Union (established in 1824), exercised a good deal of power in the regulation of pay and conditions until the decline of Thames shipbuilding in the 1860s; on Tyneside, the pro-longed depression in the industry, extending from the 1820s into the 1840s, meant that the shipwrights were less successful.[20]

The degree and strength of labour organisation was one determinant of the changing social geography of shipbuild-ing as the impact of iron and steam began to be felt from the 1840s. Certainly, the new iron shipbuilding employers setting up yards in the hitherto little developed areas of Tyneside or Clydeside, outside Newcastle and Glasgow, benefited from the relative absence of shipwrights' organisa-tion outside the established yards.[21] And for the shipwrights themselves, the development of the new industry meant that though they continued to be an important group their position of dominance over most stages of construction was lost as they were dispersed across the labour force and the range of work they undertook was restricted.

Some of the shipwrights, who as a whole formed about 10 per cent of an iron shipyard labour force, were integral to the initial work of design and planning. Those who worked in the 'mould loft' turned the plans drawn up by the draughtsmen into full-scale ones marked on the loft floor and made wooden templates and moulds for the iron plates. The draughtsmen were the primary group of technical staff. Though there was never a full separation between manual and mental labour in the industry — between those in possession of technical knowledge and a conception of the vessel as a whole, and those implementing the knowledge — there was undoubtedly a partial one, particularly as the science of naval architecture and ship construction became more complex. With this the role of technical knowledge and staff became both more clearly separated from yard labour and also internally differentiated — though the extent of this by the 1870s does not seem to have been great in most yards. Alongside these groups of workers were the patternmakers, another small but highly-skilled group, who made three-dimensional wooden patterns for the use of iron moulders and platers.

The separation of these initial tasks from the main construction work was one of the most important of developments associated with the rise of iron shipbuilding. Hull construction became the preserve, in the private yards, of the metal workers, initially introduced from iron works, engineering trades or trained up from unskilled labour. Those who dominated here were the boilermakers — the angle-iron smiths, platers and riveters — who with their assistants, helpers and labourers and the holders-up and caulkers constituted 50–60 per cent of the labour force. It was they who translated the plans and models into the reality of the ship, from the bending and shaping of angles to the preparation and fitting of plates for attachment to the ribbing by riveting and the caulking of the vessel to make it watertight. Alongside these workers were the drillers or hole-cutters who prepared the plates for rivets, blacksmiths who produced various, mainly small, metal components, and those shipwrights who engaged in the complex tasks of laying the keel on the blocks in the slipways.

As ships grew in size and complexity so too did the range of trades engaged in fitting and finishing. Part of this work remained the preserve of the joiners, forming about 9 per cent of labour, who became, like the shipwrights, one trade among many in the new industry. But the most important new group at this stage were the highly-skilled engineers working in the marine engine shops that were either part of shipbuilding establishments or in the specialist firms to be found on all the major rivers. Besides them were a range of other small fitting and finishing trades, like painters, plumbers and a variety of metal-working trades like brass and copper finishers or foundrymen.[22]

The use of new materials and building methods entailed a greater use of technical equipment and facilities than had existed in the wooden yards. But what is striking is the extent to which the industry remained based upon hard, physical labour and the skills, judgement and experience of workers and, concomitantly, tools rather than machines — i.e. equipment that was the instrument of labour rather than vice versa. For once the techniques of iron building had been established there was little technical change that radically affected, or threatened, the position of the various groups of workers, at least until the later 1870s and the subsequent use of steel. In part this was due less to the total absence of machines than to the difficulties of their widespread use. Thus, for example, machines were already available in the 1850s that could bend, shape and shear plates: but they were not widely deployed, primarily because they could not adequately cope with the technical problems of shaping plates for the curved parts of the ship. Indeed, that job remained the high art of the plater: his ability to manipulate hot metal to the right curvature was seen by contemporaries as exemplifying his skills even though it was also recognised it would be highly desirable to substitute machines wherever possible.[23] Similarly, riveting machines were available and it was known that very considerable increases in productivity could be achieved through their use. However, such machines could only be used in the easiest of circumstances: they could not be used in most of the curved parts of the ship or in awkward spaces. What remained pre-eminent was the worker and his tools and the interaction of workers rather than machines.[24]

Yet if manual labour and skills remained at a premium as they had done in wooden building, there can be little question of the magnitude of difference between the two industries, a difference which derived only in part from the use of new materials and techniques. Critically, it lay in the altering character of the relationship between employers and workers. In wooden building the compulsion to labour for an employer had been primarily economic. But in iron shipbuilding the compulsion deriving from dependence upon the wage was reinforced by a 'technical' one: the nature of the assembly process as a more highly fragmented one in which the activities of a large number of different trades were combined meant that no single group of workers could control the work itself to anything like the extent that the shipwrights had.

But, of course, the technical and economic aspects of this are not effectively separable: the basis for the more direct intervention of employers into the day-to-day processes of assembly was their increasing control over labour. One key aspect of this was the change in the character of the wage as it became more clearly a payment for the expenditure of a given amount of labour-power of particular workers, mea-sured either by time, as for most unskilled or 'semi-skilled', or by piece, as for most of the skilled. The form of payment was, of course, far from unimportant: piecework, based upon calculations of output for more or less standard tasks, facilitated and was informed by the attempt by employers to impose more direct control over the pace and quality of work, though it could never be a wholly successful means of achieving these ends.[25]

On the other side, the employers' ability to take more direct control derived, in part, from the heightened scale of capital involved in iron shipbuilding and from the assimila-tion of the overall coordination of the more fragmented and technically sophisticated assembly process to management. In some of the largest of firms this went hand-in-hand with a relative separation of ownership and control, particularly where a company became a joint-stock concern, as happened at Palmer's of Jarrow in the mid-1860s. But the typical firm remained the family-owned concern although this by no means precluded the development of a more specialised and hierarchically organised management structure involving foremen, departmental and general managers.[26]

The management of the yards did not, however, rest entirely with the managers. To an important extent, immediate authority in the implementation of tasks was invested in some key groups of workers. In particular, the boilermakers worked with labourers or assistants under their direction in an organisation of labour that bore some similarities to the gang system in wooden shipbuilding. The angle-iron smiths worked in groups of up to eight with their strikers; the platers usually worked in teams of six or seven, the specialised tasks of plating being divided between them, and each plater working with an average of four helpers. The riveters also had assistants, two riveters working with one holder-on and one or two rivet boys at the rivet hearth.

Within this system, the skilled had considerable authority over the labourers which derived, in part, from their directive role in the work itself. But it also derived from the form of wage-payment. This is most clearly seen in the relationship of the platers to their helpers. The team of platers worked for a cooperative contract negotiated with the employers, were paid on piece rates and, in turn, paid their helpers on time rates. As such there were some similarities between this and that system of subcontracting which had existed in wooden shipbuilding and was practised in a number of nineteenth-century industries.[27] But what distinguished the two was, first, that the platers were not selling finished goods to the employer but labour-power for which they received a wage, and second, that though the platers paid the helpers the rates were fixed between the employers and the helpers. Under subcontracting, wage rates were fixed independently of the capitalist putting out the contract.

For the employers the great advantage of this mode of labour organisation was that it conferred upon the platers some of the tasks of labour management while it also potentially minimised the amount of time lost through slacking and maximised the incentive to sustain high-quality work. For the workers the consequences were contradictory. On the one hand, the platers could, and often did, force the pace so as to be able to take time off — 'St Tuesday' as well as 'St Monday' was adhered to, at least in good times — while being able to get particularly high earnings when regularly employed. On the other hand, the helpers were forced to work hard for part of the week and were then laid off, with the result that, though they were by no means without skills, they were among the lowest paid workers in the yards.

Antagonism between the platers and helpers began to emerge openly in the early 1870s and focused on the system of wage payment. From about 1872, a number of shipyards on the north-east coast, beginning with one of the Hartlepool yards, began to introduce the 'corner system' under which the helpers were paid on piece rates rather than on time, receiving so much per plate handled. As we shall see, the system did not last long but the conflicts that developed over it revealed much of the nature of the relationship between helpers and platers. Not least, they showed that the platers did not enjoy untrammelled power over the labourers. As with the related boilermaking trades — the smiths with their strikers, the riveters with their holders-up — the nature of the work itself demanded a degree of cooperation and the establishment of a rapport between the platers and helpers, not least because some of the tasks in which all these groups were engaged entailed the manipulation of hot metal. What perhaps made the difference between their relationship and those other groups, and helps to explain why conflicts between them were much more pronounced, was that the ratio of helpers to platers was higher than that of strikers to smiths or holders-on to riveters, and that the helpers were beginning to organise.[28]

There can be little question that the highly skilled workers continued to enjoy considerable authority within the yards which, in important ways, gave them a stake in the maintenance of the existing division of labour. But this did not lead them to a passive acquiescence in the overall authority and control of the employers. Indeed, it was precisely among these workers that the best organised trade unions developed. But that these did so needs to be seen in relation to another feature of the division of labour in the industry.

So far it has been tacitly assumed that the division of labour was a natural outcome of the development of new materials and new building techniques. It is evident however that this was by no means the case: there was no *a priori* reason why iron shipbuilding could not have developed by multiplying the numbers of shipwrights at the same time as sub-dividing the work into more detailed stages. This, after all, happened in the government dockyards where the shipwrights made the transition from working in wood to working in iron and where the boilermakers were unable to gain the same kind of foothold as in the private yards. But this was not because methods of working were radically different: rather it arose out of the manner of control which the government, as employer, could exercise, restricting, as it did, both trade union growth and the elaboration of rigid demarcation between trades.[29] That the development of the division of labour was so different in private yards must largely be attributed to the peristent refusal of wooden 'wrights to work in iron, as they regarded the boilermakers as an 'inferior' class of tradesmen. Their inability to keep iron men out was compounded by a number of other factors, though these varied from area to area. On Tyneside, for example, virtually all the companies that began to undertake iron shipbuilding in the early 1850s were wholly new to the industry and established themselves by bringing the workers from other industries and other areas. At the same time, the shipwrights were weakened both by their unsuccessful strikes of 1841 and 1851 and, in the later 1850s, by the depression in the market for wooden ships.[30] In turn, the entry of the boilermakers into the industry, and their organisation into what was to become the most powerful of shipyard trade unions, enabled them to fight against the encroachments of shipwrights on their work. From the 1860s it was becoming increasingly evident to shipwrights that they must come to terms with the new materials but their ability to do so was handicapped by the strength of the boilermakers, whose union were imposing fines on any members who worked with shipwrights on iron work.[31]

That the boilermaking trades were able to enter and dominate the central constructive areas of shipbuilding was the basis for building a position of considerable power, both in the organisation of work and in collective bargaining with the employers. By the 1870s boilermaking was probably the best organised trade in the country and certainly the best organised in this industry.[32] But again this should not be seen as a natural outcome of the division of labour and these workers' position within it. Rather it was a position built up through the establishment of an effective trade union able to defend its members by a number of means. It entailed, first, the maintenance of a five-year apprenticeship, preferably rooted in the recruitment of the sons or other kin of existing boilermakers, as the means of maintaining the scarcity of skills. Second, it meant attending to the organisation of the unskilled where this might threaten working conditions or pay. In the case of the helpers this led to attacks by the Boilermakers' Society upon their attempts to organise in the 1870s; in that of the caulkers, it entailed their admission to the Society in the early 1870s, as some of the members were able to show that the caulkers had a legitimate right to be considered a section of the trade and, moreover, did not constitute a threat to other trades.[30] Third, it involved the attempt to establish standard rates of pay and conditions, at the very least on a district basis, which necessarily involved achieving recognition from the employers and the creation of an effective machinery of collective bargaining. Neither of these aims had been wholly achieved by 1880 although they very nearly had in some areas. Finally, attention had to be paid to the claims and threats of other groups of skilled workers and their unions in the yards. The fight with the shipwrights was central here but it was by no means the only one. In 1865–6 for example, the Boilermakers' Society fought the Amalgamated Society of Engineers over the encroachments of enginesmiths onto the work of angle-iron smiths; it also fought, a decade later, the Associated

Blacksmiths over the employment of a blacksmith to do angle-iron work.[34]

The Boilermakers' Society was distinguished not only by its size as the largest of the metal-working unions in the yards, but also by its ability to sustain organisation. In this industry of severe cyclical fluctuations insecurity was a chronic and ever-present problem for all workers. For those without the means of maintaining entry to the trade or of effectively regulating the conditions of work the problem of sustaining any organisation proved insuperable in this period. The platers' helpers, the largest single group of labourers, attempted to organise as a particular occupational group in the 1870s, but their unions remained localised and were fairly short-lived.[35] Where the unskilled could enter trade unionism they could do so effectively where a skilled union would take them, as happened with some blacksmiths' societies.[36]

But if it was the skilled who formed unions, or had any reasonable chance of forming and maintaining them, the extent to which they did so was uneven and affected by the competing claims of other trades. Among the engineering and metal-working trades a number of societies competed for members and for the means of preserving the regulation of the trade — the ASE, the Steam Engine-Makers, the Patternmakers, the Blacksmiths and the Boilermakers all developed varying degrees of strength in the yards.[37] And the proliferation of unions in the yards led to considerable problems: relations between them frequently displayed a lack of cooperation that is one element in explaining why an industry as well organised as this was by 1880 was also one in which machinery of collective bargaining involving all the trades was relatively difficult to establish. The great majority of disputes in the industry were those between one trade and an employer or group of employers on a river. Where the whole of the shipyard labour force was in dispute with the employers the conduct of it was often marked by a notable lack of cooperation between the trades, with each trade tending to negotiate separately, as happened, for example, in a major dispute on the Tyne early in 1875.[38]

At the same time, it would be wrong to exaggerate the

extent of conflict between the skilled groups in this period and to see those of the 1880s and 1890s as simply continuing an established pattern. The earlier disputes had the form of sporadic and often scarcely noticed battles; the later ones were rather more like sustained warfare. This was due, in part, to the generally favourable conditions in which workers operated in the third quarter of the nineteenth century. In the more severe technical and economic climate of the late nineteenth century, conflicts between different groups of workers and their trade unions became both more widespread and bitter.

III

The shift from iron to steel as the main hull construction material took place in the early 1880s as a result of rapid reductions in steel prices and improvements in its quality. This did not involve such enormous transformations in the industry as the earlier shift from wood to iron had done, for the basic assembly process remained the same, as did the size of enterprises, their management structures and their regional location. But while there was no challenge to the basic principle that boilermakers should control the handling of metal in hull construction, they did lose some ground during the changeover to steel.

This was most serious in the case of the platers, for the greater ease of manipulation of steel in the cold state finally eliminated the handling of heated metal plates, and turned plating into an occupation based entirely on shaping and bending tools. This was not mechanisation in the full sense, as a great deal of technical aptitude and initiative was still required, and the tools were not effectively linked into a unified process under management control, but it did cause a significant decrease in the level of skill of both the platers and their helpers.[39] The skilled men's previously unchallenged technical direction of the squad came under pressure, especially as their helpers began to claim a right to some of the machine work in order to prevent their own reduction to mere haulage workers. In response the platers took firm

steps to restrict their helpers to simple manual work as much
as possible, and to remove any influence they had had over
the pace of work by the abolition of the practice of sharing
the piece earnings of the squad under the 'corner system',
and by refusing to pay them on any basis other than time
rates.[40] Enforcing this strategy produced a series of noto-
rious sectional disputes between the two groups in the early
1880s, most notably in Sunderland where the skilled men
took strike action to force the employers to accept it. During
the dispute the platers' fears were fully confimed as the
shipwrights and other tradesmen began to show the helpers
how to use the machines and effectively to replace the
skilled men. But the employers were unable to treat this as a
serious strategy because the firm stance of the boilermakers'
union as a whole reminded them of their dependence on its
other sections to maintain production: they conceded to the
platers' demands and imported 700 blackleg helpers from
outside the district to replace the intransigent local men.[41]

In the face of this decisive precedent the 'corner system'
gradually crumbled elsewhere and the platers were able to
consolidate a relatively favourable degree of control over
the new processes, albeit at a lower level of all-round skill
and with a much deteriorated relationship with their helpers.
Because of these frictions involved in the adaptation to steel
combined with the financial unviability of admitting such
large numbers of lower-paid men into the union, the platers'
helpers continued to be firmly excluded from the Boilermakers' Society. Over the next decade they were chronically in
dispute with it over their loss of independence, their
increasing insecurity of employment, and the low levels of
time wages they received from the platers (largely because
of these pieceworkers' irregular timekeeping customs).[42]
However, the platers' helpers' ability to organise themselves
re-emerged during the 1889 boom, and proved to be
permanent. Indeed they established one of the strongest
unions of the unskilled in this period, and were able to
engage in collective bargaining with the employers and to
improve their conditions and earnings by gaining direct
employment from the yard office.[43] While the platers had
initially resisted the demands of the new helpers' organisa-

tions for higher wages from them, the switch in the tactics of the unskilled towards demands on the employers won them the passive support of the boilermakers' union whose members by now preferred to have absolutely no responsibilities as 'masters' over the unskilled.

The other major threat posed by steel was to the riveters, for its greater strength per unit of weight made it practical to use significantly larger hull plates and thus simply and immediately reduce the amount of riveting by as much as 25 per cent.[44] However, the rapid expansion of output and the increasing size of individual vessels more than offset this threat of labour displacement, and one other feature of steel work actually strengthened the riveters' position. Because of its greater malleability it was possible to form complex beams and angles for the internal supportive structure of ships from only one steel beam, whereas using iron such items had to be made from several straight pieces riveted together. This increased the amount of work available for the angle-iron smiths, who were able to consolidate their position as the most skilled and best paid among the boilermakers.[44] And, while it completely displaced riveters from this area of production, it paradoxically also reduced the visible threat of machinery to their position by reducing the proportion of total riveting done by machine.

At this time the only effective riveting machines were hydraulic ones, which were so immensely powerful that they could only be used on jobs for which it was not necessary to countersink the rivet flush with the steel surface: effective countersinking would have required some contact between the jaws of the machine and the steel components being assembled and this would have resulted in their distortion or fracture. Consequently, hydraulic machinery was not suitable as a substitute for hand work on the outside shell plates of the vessel which required countersinking in order to minimise water resistance and which comprised the vast bulk of riveting work. Machine riveting was therefore confined to heavy keel and beam riveting on the berth and to the pre-assembly of structural components at a stationary machine, and this, especially after the displacement of much pre-assembly riveting by the bending of one steel beam, amounted to only around 5 per cent of total riveting on even the biggest vessels.[46]

Employers remained keen to reduce their enormous bills for riveting labour, but were restricted to changes in design such as the introduction of larger plate sizes, and the overlapping of plates in increasingly efficient ways. This combination of the absence of a threat to the nature of the skilled work from extensive mechanisation with these less obvious though none the less real pressures, led to a strengthening of riveters' defences against the employers' encroachments by the organisation of all the adult members of the squad into the boilermakers' union. Skill differentials had always been smaller between the riveters and the holders-on than those between platers and their helpers even after the introduction of plating machinery, but the ratio of unskilled to skilled was lower in the riveting squad and there was a strong tendency for its members to stick together and frequently to be relatives. It was therefore both desirable and financially practicable to open the union formally to these helpers in 1882, and then forcefully put this decision into practice during the depression of the mid-1890s.[47]

This strengthening of organisation of the riveting section, combined with the organisation of other interdependent structural steel workers in the same union, consolidated the position of riveters in the face of the more severe threat of the pneumatic riveting machines which first appeared in the 1890s. These tools which were much lighter than the hydraulic machines, were constructed as hammers which could effectively simulate hand work, even on jobs requiring countersinking. Not only could they be operated efficiently by only one man instead of the customary two-man hand riveting squad, there was no intrinsic reason why this operator should have been a skilled riveter.[48] However, the enormous threat of this machinery to the position of the skilled men was relatively easily contained by the organisational strength of the boilermakers' society which was able to enforce a rule that *full* squads of skilled men be employed wherever the pneumatic tools were introduced. Thus, though some of the largest liner and warship building firms went ahead with the capital expenditure on this equipment, the only benefit was in terms of reductions in the rates of pay to the skilled men, and even this was restricted by the union to specified deductions from the existing rates for hand work. Simultaneously, pneumatic tools were found to be even more effective for the less skilled and more repetitive work of the caulkers, and could have reduced their numbers by 25 per cent. However, this group benefited from its earlier admission (in 1877) to the Boilermakers' Society, which was once again able to restrict the operation of the tools strictly to its own members, to control the displacement of labour, and to link the new rates of pay to the established rates for hand work. Thus in both cases the impact of the introduction of pneumatic tools in the 1900s was firmly contained within the already established trade controls and collective bargaining procedures carefully built up by the union over the preceding generation.[49]

The main groups of hull construction workers, including the blacksmiths and the shipwrights who cannot be dealt with in detail here, while involved in significant conflicts over such issues as numbers of apprentices, the extension of piece rates and systematic overtime, and of course the level of wage rates, did not experience any substantial transformation of their positions in the division of labour as a result of mechanisation between 1880 and 1914. This was less true among the outfitting trades where, by contrast, machinery was being introduced, partly as a result of pressures in sectors outside shipbuilding in which these groups also worked. The engineers, dealt with in more detail in chapter 6 of this book, were faced with a major threat from machinery and reorganisation in the 1890s. While in the specialist marine engine works with their higher quality and absence of opportunities for repetition production there was little mechanisation, there *was* a marked increase in employers' pressure over numbers of apprentices, and the introduction of piece rates and overtime. This was much harder to resist given the weakening of the occupation's position nationally, and the problems the union faced in organising effective resistance.

A significant threat was also posed to both the plumbers

and the joiners by new machinery, though in both cases it was more successfully contained than in engineering, being eventually restricted to the mass manufacturing of semi-processed components in separate shops and enterprises without significantly encroaching on the preserves of the skilled assembly workers. In the case of the plumbers this seems to have been due to a combination of a high level of skill in assembly work, a lack of decisive pressure on the trade from the employers, and the increasing employment opportunities available in an expanding industry in which the outfitting work was simultaneously becoming more complex.[50] While these latter changes in the industry also benefited the joiners, in their case the strengthening of union district and national organisation played a far more vital part, for on this basis they were able to restrict many of the machines to the wood preparation shops, to abolish piece rates and to maintain customary practices and standard time rates in the joiners' shop itself.[51] The plumbers were unusually fortunate, perhaps because many employers were using them to encroach on the boundaries of engineers' work on board ship, but with this one exception the outcome for occupations faced with mechanisation or significant changes in the division of labour was by this period closely related to their ability to form effective, usually more centralised, national trade union organisations which could enforce general and decisive policies.

This was of course pre-eminently the case with the Boilermakers' Society itself, renowned as one of the most powerful skilled workers' unions in the country. By this period the boilermakers, though not threatened with extensive displacement by machinery, were no longer a highly skilled group in terms of the work they performed. Their degree of all-round aptitude was diminished not only by the subdivision of the trade into four separate sections (angle-iron smiths, platers, riveters, and caulkers) between which it was rare to move, but also by further specialisation within each section.[52] Thus, for example, platers' squads by this period specialised in light, heavy and ordinary work, while within each squad the individual workers tended to specialise in particular tasks. Not only were they restricted to increasingly narrow spheres of operation, but few boiler-makers in the late nineteenth century were manipulating hot metal. Traditionally it was the intuitive judgement arising from long practice in the firing of furnaces, the heating of the metal and the experience of when, where and how to strike it which had been the basis of metal workers' skills.[53] While this remained true of the separately organised blacksmiths, among the boilermakers only the angle-iron smiths were still involved in the skilled manipulation of heated bars. The riveters did still use hot rivets but this only required very basic levels of judgement and a few technical skills, while the platers now used large machine tools to manipulate *cold* steel.

However, at the same time as the skill content of their work had declined the boilermakers had managed to construct an almost impregnable defensive position through the formal controls of trade union organisation. Not only did they have the advantage of organising several interdependent occupational groups which were central to production (having admitted the semi-skilled caulkers and holders on in 1877 and 1882 respectively), they had, by the 1890s, virtually a closed shop in each group. Short of a complete reorganisation of ship production, such as that which was attempted after the second world war, it was almost impossible to alter significantly the working conditions of any section of the union for fear of provoking a complete close down of the yard or even the industry. This impeded the introduction of pneumatic riveting tools and reduced the impact of plating machinery by placing strict limitations on the extension of the labour force, on the cutting of rates, and on the speeding-up of the pace of work. Apart from the regulation of machinery the union was also able to evolve and enforce centralised decisions on the restriction of the numbers of apprentices employed, so that, while the yard owners were frequently able to have their way during slack periods, the union never suffered an irreversible defeat on this issue. It was similarly successful in eliminating the submission of individual squad contracts for jobs (a practice which had led to competitive undercutting of rates), an imposing *collective* bargaining throughout the industry and in gradually equalising rates of pay across districts.[54]

As well as struggling against the employers to establish institutional controls over the size of the occupational labour market and the framework of collective bargaining, the Boilermakers' Society continued to struggle against other groups of workers over the definition and extension of the boundaries of its members' work. One part of this strategy involved varying policies towards the unskilled helpers of its various sections, either admitting them as in the case of the holders-on and the rivet boys, or taking strong steps to exclude them, as in the case of the platers' helpers. Meanwhile, the second part of the strategy towards groups outside the union involved attitudes to workers already substantially organised on an independent basis. In the case of the drillers, the boilermakers were initially hostile to the strengthening of their organisations in the 1889 boom. Though they did eventually recognise the existence of the United Drillers' Society, they still claimed drillers' work for their own members and persistently turned down the offers of the UDS to dissolve itself into the Boilermakers' Society.[55] At the same time, they were involved in severe demarcation disputes with the shipwrights' union over plate handling work on the building berths. Though these were increasingly resolved by joint union committees and the two societies established a more harmonious relationship, the boilermakers prevented closer unity in this case by demanding complete absorption of the shipwrights' union, which was prepared only for federation.[56]

Clearly then, by this period the issue of on-the-job demarcation of work was intimately bound up with the increasing strength of the institutions of labour, formed in the first instance to defend skilled workers against the encroachments of the employers. This pattern was equally evident outside the central hull construction trades, with the engineers, for example, putting increasing pressure on such neighbouring occupations as the patternmakers, the steam-engine makers and the plumbers. While this was probably a result of the growing pressure from the employers which they were themselves subject to, it was a self-defeating response as these neighbouring groups all refused to support the ASE during the 1897 lock-out.[57] In some contrast, the increasing pressure by the joiners on their neighbours the shipwrights, was probably related to their success in resisting employers' pressures and their desire to expand from a position of increased organisational power.

Even the emergence of the Federation of Engineering and Shipbuilding Trades by no means overcame the vexed question of sectional conflicts, on many occasions functioning merely as a forum for their expression, and sometimes even intensifying them by its very existence. This was especially true of the relationship between the boilermakers and the engineers whose long-standing rivalry was continually reinforced by their undisputed predominance and isolationist policies in the shipbuilding and engineering industries respectively, in each of which an important minority of their rival's members were also employed. As a result of this rivalry the ASE refused to join the Federation until 1905, after which its presence may have strengthened that body, but simultaneously challenged the boilermakers' hegemony over it for the first time and accordingly produced many internal disputes over policy. In one ironically symbolic instance in 1912, the Federation reached a new agreement with the employers' associations over the procedure for resolving demarcation disputes, a move which had been strongly resisted by the boilermakers, and which resulted in them bitterly accusing the engineers of packing Federation meetings in order to bulldoze the policy through.[58]

In this period, then, the reorganisation of work (sometimes involving the introduction of machinery) and the strengthening of trade union institutions were both evident in the shipbuilding industry and were clearly closely interrelated. However, pressures on the established division of labour by no means automatically led to an increasing homogeneity among workers in the industry, nor did increasing levels of unionisation automatically produce greater unity and cooperation. Indeed, both developments seem frequently to have produced quite the opposite effects of those customarily associated with them.

IV

While the major questions of culture and politics raised by the literature on the labour aristocracy have been beyond the scope of this paper, it seems that it would be hard to find a clear economic basis for such a stratum in the shipbuilding industry. We cannot draw a simple contrast between the skilled and the unskilled when looking at levels of technical competence. Helpers, for example, had always had varying degrees of task-specific aptitude and, as the impact of economic and technical change was felt in the course of the century, many of the tradesmen had surprisingly low levels of all-round skill and on-the-job initiative. Similarly, there was no simple contrast between forms of trade unionism, as skilled workers established a variety of types of organisation and the helpers' organisations pursued occupational rather than industrial union strategies. Equally, few even among the most highly skilled had real security of employment and their incomes were consequently lower than their rates of wages might suggest: at the same time there are indications that some of the unskilled had relatively higher incomes than might be expected. In short, there was no simple correlation between the three hierarchies of skill, organisation and earnings. Rather than seeing this industry as containing a coherent upper stratum of skilled workers, it should be seen as one characterised by considerable levels of sectional fragmentation.

One of the major themes of this paper has been the relationship between technical constraints and union strategies, particularly in the case of the replacement of the shipwrights by the boilermakers and the consolidation of a strong defensive position by the latter group. It seems that in periods of major innovation and reorganisation, such as that involved in the transition from wood to iron, the constraints of technology are loosened and industries are relatively open to the emergence of new occupational configurations. Thus in the iron shipbuilding industry other divisions of labour were not only theoretically possible but demonstrated in practice in the government dockyards. The outcome was therefore not technically determined, but emerged as a

Table 5.1: The size of the main occupational groups in the shipbuilding industry, Harland & Wolff, Belfast, 1892.

Shipyards		
Shipwrights	441	9.6%
Riggers, etc.	66	1.4%
Patternmakers		—
Blacksmiths	90	2.0%
Finishers	10	0.2%
Strikers	117	2.5%
Platers	360	7.8%
Angle-iron smiths		—
Platers' helpers	846	18.4%
Riveters	400	8.7%
Holders-on	177	3.9%
Caulkers	181	3.9%
Drillers	189	4.1%
Engineers	139	3.0%
Fitters' helpers	107	2.3%
Plumbers	72	1.6%
Electricians		—
Joiners	381	8.3%
Painters	53	1.2%
Red leaders		—
Miscellaneous	180	3.9%
Labourers	486	10.6%
Boys	294	6.4%
Total	4589	100%

Engine works		
Patternmakers	48	1.9%
Joiners	43	1.7%
Patternmakers' assistants	9	0.4%
Iron-moulders	157	6.3%
Iron-moulders' helpers	185	7.5%
Brass-moulders	23	0.9%
Brass-moulders' helpers	20	0.8%
Turners	120	4.8%
Fitters	436	17.6%
Fitters' helpers	448	18.1%
Machinemen	129	5.2%
Boilermakers	234	9.5%
Boilermakers' helpers and boys	424	17.1%
Blacksmiths	40	1.6%
Blacksmiths' strikers	61	2.5%
Miscellaneous	7	0.3%
Labourers	91	3.7%
Total	2475	100%

Source: Evidence sent by Sir Edward Harland to the Webbs referring to week ending 9 August 1892, in Webb Trade Union Collection, E.A. XV(1), pp.11–12.

result of the unwillingness, and later the inability, of the shipwrights' societies to develop an effective strategy for retaining the main tasks of ship construction when confronted with the intrusion of the boilermakers. After this major transition was completed, the relationships between technology, occupations and trade unions became relatively stable and union policies had more of the appearance of tactical responses to economic interests and technical determinants. However, even in this period, the variations in the policies of the Boilermakers' Society towards different groups of less skilled assistants and different groups of already organised workers indicates the continuing importance of developing union strategies even within a relatively closed industrial structure. The Boilermakers' Society seems, in one respect, to have been a typical craft union, but by the late nineteenth century was including workers who, in technical terms, could be considered semi-skilled. Trade unionism, then, was not simply a reflex response to pre-given situations but rather a matter of conscious organisation and of social power.

NOTES

Our thanks to John Lovell, Joseph Melling and Steven Tolliday for comments on an earlier draft. Thanks also to the President of the Amalgamated Society of Boilermakers, Shipwrights, Blacksmiths and Structural Workers (now amalgamated with the GMWU) for permission to consult the Society's records.

1. E.J. Hobsbawm, 'The Labour Aristocracy in Nineteenth-century Britain', in *Labouring Men* (London,1964), pp.272–315, esp. pp.284–8. H. Pelling, 'The Concept of the Labour Aristocracy', in *Popular Politics and Society in Late Victorian Britain* (London, 1968) pp.37–61 (quotation from p.60).
2. *Royal Commission on Labour*, Group A, Third Report, c.6894, Parliamentary Papers 1893–4, XXXII, q.20, 802. A quotation which took on a life of its own after being included in E.J. Hobsbawm, *Labour's Turning Point* (London, 1948), pp.4–5.
3. S. Pollard and P. Robertson, *The British Shipbuilding Industry, 1870–1914* (Cambridge, Mass., 1979), pp.9–18.
4. Ibid., pp.70–72; S. Pollard, 'The Economic History of British Shipbuilding 1870–1914', PhD, University of London, 1950, pp.51–5.

5. Alastair Reid, 'The Division of Labour in the British Shipbuilding Industry, 1880–1920', PhD, University of Cambridge, 1980, pp.11–14.

6. S. Pollard, 'British and World Shipbuilding, 1890–1914: A Study in Comparative Costs', *Journal of Economic History*, 17 (1957), 426–44; Reid, *op. cit.*, pp.32–9.

7. *R[oyal] C[ommission] on Labour*, Parliamentary Papers, 1893–4, c.6894–VII, XXXII, qq.20, 676–82.

8. Ibid., qq.20, 683.

9. Reid, *op. cit.*, pp.39–47.

10. Pollard and Robertson, *op. cit.*, pp.29, 152.

11. *Report on an Enquiry by the Board of Trade into the Earnings and Hours of Labour of the Workpeople of the United Kingdom. VI. Metal, Engineering and Shipbuilding Trades in 1906*, c.5814, Parliamentary Papers 1911, LXXXVIII, pp.64–5, 107–8.

12. For a rough indicator of fluctuations of wages, see the movements of rates in A.L. Bowley and G.H. Wood, 'The Statistics of Wages in the United Kingdom during the Nineteenth Century. Part XIV. Engineering and Shipbuilding. E. Averages, Index Numbers, and General Results', in *Journal of the Royal Statistical Society*, LXIX (1906), pp.148–96, esp. pp.176–7.

13. S. Price, 'Riveters' earnings in Clyde shipbuilding, 1889–1913', *Scottish Economic and Social History* 2 (1981), pp.42–65; on these issues see also the interesting comments in E. Roberts, 'Working-class Standards of Living in Barrow and Lancaster, 1890–1914', *Economic History Review*, XXX (1977), pp.306–21.

14. Reid, *op. cit.*, pp.47–50, 139–42.

15. See, e.g., Keith Burgess, *The Origins of British Industrial Relations* (London, 1975), chs 1 and 4.

16. Pollard and Robertson, *op. cit.*, p.152.

17. E.P. Thompson and E. Yeo (eds), *The Unknown Mayhew: Selections from the 'Morning Chronicle' 1849–1850* (London, 1971), pp.399–413.

18. *S[elect] C[ommittee] on Navigational Laws*, Parliamentary Papers 1847 (678)X, qq.8003 *et seq.*; S. Pollard, 'The Decline of Shipbuilding on the Thames', *Economic History Review*, 3 (1950–1), 72–4.

19. I. Prothero, *Artisans and Politics in Early Nineteenth-Century London* (Folkestone, 1979), pp.250, 381 nn.36–7; *Select Committee on Manufactures, Commerce and Shipping*, Parliamentary Papers 1833 (690)VI, qq.6658, 6749, 6755–9.

20. Prothero, *op. cit.*, Pollard, 'Decline of Shipbuilding on the Thames', 74–5; Philip Rathbone, 'An Account of Shipwrights' Trades' Societies in Liverpool, the Tyne, and other ports', National Association for the Promotion of Social Science, *Trades' Societies and Strikes* (London, 1860), 479–520; K. McClelland, 'Skilled Workers on Tyneside, c.1850–1880', PhD University of Birmingham, forthcoming.

21. Pollard and Robertson, *op. cit.*, pp.49–58.

22. There is no description of the division of labour in this period comparable with those for the post-1880 period of A.C. Holms, *Practical Shipbuilding* (2 vols, London, 1904), or Reid's thesis (*op. cit.*); but there is much of interest in J. Grantham, *Iron Shipbuilding* (London, 1858); E.J. Reed, *Shipbuilding in Iron and Steel* (London, 1869); D. Pollock, *Modern Shipbuilding and the Men Engaged In It* (London, 1884); a guide to occupations is the 'Glossary of Terms' in *Royal Commission on Labour*: Parliamentary Papers 1893–4, c.6894-X, pp.97–134.
23. Grantham, *op. cit.*, p.56.
24. See esp. Reed, *op. cit.*, pp.327, 340–5.
25. See, e.g., William Denny, *The Worth of Wages* (Dumbarton, 1876); A.B. Bruce, *The Life of William Denny* (2nd edn, London, 1889), pp.53, 75–78, 111–16.
26. Although concentrating on post-1880, Joseph Melling, ' "Non-Commissioned Officers": British employers and their supervisory workers, 1880–1920', *Social History*, 5 (1980), 183–221, is helpful.
27. For a broadly drawn distinction between subcontracting and helper systems, see Howard Gospel, 'The Development of Management Organization in Industrial Relations. A Historical Perspective' in K. Thurley and S. Wood (eds) *Industrial Relations and Management Strategy* (Cambridge, 1983), pp.94–97.
28. For platers, helpers and the corner system see, *inter alia*, Royal Commission on Trade Unions, Parliamentary Papers 1867–8, 3980-V, XXXIX, qq.17,709–12; United Society of Boilermakers and Iron Shipbuilders, *Monthly Reports*, January 1874, July 1877; J. Lynch, 'Skilled and Unskilled Labour in the Shipbuilding Trade', *Report of the Industrial Remuneration Conference* (1885), 114–18; Webb Trade Union Collection, A.XXXIII, ff.212, *et seq.*; D.C. Cummings, *Historical Survey of the United Society of Boilermakers and Iron and Steel Ship Builders* (Newcastle upon Tyne, 1905), p.87.
29. John Field, 'Bourgeois Portsmouth: Social Relations in a Victorian Dockyard Town, 1815–1875', PhD, University of Warwick, 1980, ch.7.
30. McClelland, *op. cit.*
31. An interesting account of the friction between them is in the *Newcastle Chronicle*, 26 July 1862.
32. George Howell, 'Trade Unions: Their Nature, Character, and Work', *Fraser's Magazine*, new series, 19 (1879), 24.
33. Cummings, *op. cit.*, pp.60, *et seq.*
34. Webb Trade Union Collection, A.XXXIII, f.111; Angela Tuckett, *The Blacksmith's History* (London, 1974), pp.83–4.
35. Webb Trade Union Collection, A.XXXIII, ff.212, *et seq.*
36. E.g., The Sons of Vulcan United Society of Smiths and Strikers (establ. 1865), later re-formed in 1886 as the important United Kingdom Amalgamated Society of Smiths and Strikers. Tuckett, *op cit.*, pp.75–6.
37. Paul Robertson, 'Demarcation Disputes in British Shipbuilding

before 1914', *International Review of Social History* 20 (1975), 220–35.
38. The Tyneside dispute can be traced through, esp., the *Newcastle Weekly Chronicle*, January-February 1875.
39. J.R. Hume, 'Shipbuilding Machine Tools' in J. Butt and J.T. Ward, (eds), *Scottish Themes* (Edinburgh, 1976); P. de Rousiers, *The Labour Question in Britain* (London, 1896), p.252.
40. *Royal Commission on Labour*, *op. cit.*, qq.20, 668–71, 20, 801–11.
41. Cummings, *op. cit.*, pp.111–12.
42. Lynch, *loc. cit.*
43. *Royal Commission on Labour*, *op. cit.*, qq.20, 540–8.
44. D. Pollock, *The Shipbuilding Industry* (London, 1905), pp.97–9.
45. Ibid., pp.90–3.
46. Holms, *op. cit.*, pp.300–1.
47. Cummings, *op. cit.*, pp.112, 130; Webb Trade Union Collection, E.A. XV(1), p.193.
48. Pollock, *op. cit.*, (1905), pp.107–12.
49. USBISS, *Rule-Book* (1912), p.160; Holms, *op. cit.*, p. 306; Reid, *op. cit.*, pp.126–7, 135–6.
50. *Royal Commission on Labour*, *op. cit.*, qq.23, 378–93.
51. Ibid., qq.22, 073–094.
52. C. Booth (ed.), *Life and Labour of the People in London* (London, 1895), vol.V, p.324.
53. *Royal Commission on Labour*, *op. cit.*, qq.20, 883; 20, 887–8.
54. Reid, *op. cit.*, pp.136–44.
55. J.E. Mortimer, *History of the Boilermakers' Society*, vol.1 (London, 1973), pp.133–4.
56. D. Dougan, *The Shipwrights* (Newcastle upon Tyne, 1975), pp.166–8.
57. Reid, *op. cit.*, pp.91–5; 171–2; 175–6.
58. United Society of Boilermakers and Iron and Steel Shipbuilders, *Annual Report* (1912), p.vii.

6. Engineers and Compositors: A Comparison

Jonathan Zeitlin

I

The principal interpretations of skilled workers and their organisations in Victorian Britain have tended to rest on a set of largely unexamined assumptions about the relationship between economic development and the division of labour on the one hand, and a recognition of the distinctive features of the British economy on the other. Whether influenced by Marxism or by mainstream economics, historians of nineteenth-century Britain have generally accepted the view that the logic of industrial development — operating through the mechanisms of competition, concentration and technical change — tends radically to reduce the demand for skilled labour and so to subvert the basis for craft organisation, together with other sectional divisions within an increasingly homogeneous working class.

The bulk of the debate about the economic formation of the working class in nineteenth-century Britain has accordingly focused on *when* rather than *whether* these developments reached fruition. Thus the exponents of the labour aristocracy thesis in its classic form, such as Hobsbawm and Gray have argued that the specific features of the mid-Victorian economy — its pioneer character, small-scale enterprises, underdeveloped managerial techniques, and above all its dominant position in world markets — created a space within which skilled workers could carve out for themselves a privileged position of high and secure earnings based on their centrality within the labour process and buttressed by trade union organisation.[1] Leaving aside the disagreements over the social and political implications of these developments which have increasingly come to dominate the literature, the principal opposition to this view has contended that these processes were already well underway by mid-century. Thus Pelling claimed that apart from the survival of a few isolated handicraft trades, the working class had become relatively homogeneous by the third quarter of the century, while Foster and Stedman Jones, from rather different political perspectives, also questioned the reality of the skills possessed by the so-called labour aristocrats.[2]

Recent literature on the division of labour in advanced industrial societies has tended, however, to undermine these general assumptions about the impact of economic development on skill levels shared by both the proponents and opponents of the labour aristocracy thesis. Thus it has become clear that the development of the division of labour spontaneously reproduces a demand for a significant, if perhaps declining, number of workers with general skills; that the division of labour is itself strongly conditioned by worker organisation, market structures and managerial strategies; and that major technical innovations hailed as avatars of deskilling have had contradictory consequences in practice.[3]

These findings have a double implication for analyses of skilled workers and the persistence of craft regulation in Victorian and Edwardian Britain. First, the assumption that the late nineteenth century saw a rapid erosion in the position of skilled workers and in the effectiveness of craft regulation as a consequence of widespread mechanisation spurred by foreign competition needs to be re-examined in the light of the experiences of particular trades. Employers do not necessarily, or even normally, respond to intensified competition by rapid mechanisation, and such technical innovation as does occur may enhance rather than undermine the prospects for craft regulation. Hence, where technical and organisational change does result in the subversion of craft regulation, the causes need to be located within the specific circumstances of the industry, and in the outcome of struggles between skilled workers and their employers to control the conditions of their introduction. Conversely, it also becomes necessary to reconsider the preceding period of British hegemony in world markets in its own right rather than treating it as a period during which more general tendencies of capitalist development were temporarily held in abeyance. Just as mechanisation need not eventuate in the demise of craft regulation, so too there are other forces which may work to undermine the latter, and the position of skilled workers and their organisations needs to be evaluated in the light of pressures on employers to cheapen and intensify their labour within a technically static division of labour.

Engineers and compositors, the groups which form the subject of this chapter, have always figured prominently in discussions of the labour aristocracy. Thus engineering has been termed the '*locus classicus*' of the labour aristocracy, and its experience forms the basis for the archetypal pattern of mid-Victorian prosperity and *fin de siècle* decline.[4] Similarly, even Pelling is prepared to accord an aristocratic status to compositors, though Hobsbawm confines membership of the super-élite of the highest paid to the newspaper branch, and Gray notes the embattled position of the trade in Edinburgh.[5] Following the considerations adduced above, this chapter will examine the effectiveness of craft regulation during the period of relative technical stability which prevailed in each industry after mid-century, followed by an analysis of the broader confrontations over the division of labour touched off in the 1890s, together with their resolutions to 1914.

II

The principal threat to the position of the hand compositor in mid-nineteenth-century Britain came, paradoxically, from the rapid expansion of the printing industry. With the repeal of the 'taxes on knowledge' in the 1850s and the extension of popular literacy, came a vast growth in the demand for printed matter. Total male employment in the printing industry tripled between 1851 and 1891, an increase which was concentrated above all in the newspaper and periodical press.[6] While the same period saw a series of transformations in the technology of the printing press which radically increased the speed and volume of its output, the composing room remained much as it had been in the days of Caxton and Gutenberg. Until the introduction of the linotype in the 1890s, printing employers remained dependent on hand labour for the crucial operations whereby the individual moveable types were transferred into galleys and the latter made up into pages and arranged in the correct order for printing. Hence, as the industry expanded, the composing room loomed progressively larger as a bottleneck to larger and faster print-runs and as a department in which labour costs failed to decline with increased output.[7] Hence in the absence of a technically satisfactory solution to the problem of mechanical composition, printing employers found themselves forced to experiment with a variety of strategems for cheapening and intensifying hand composition within the framework of the existing division of labour.

While the physical process of typesetting was itself relatively routine, requiring principally manual dexterity and a knowledge of the location of the types in their cases, the compositor's job as a whole demanded a broader range of intellectual skills. The hand 'comp' had to be able not only to read, but also to decipher often illegible handwritten copy, to justify the lines and internal spacing following complex rules of printing styles, and to correct spelling and punctuation. The book hand might be called upon to set matter in foreign languages, or perhaps to arrange tables, while the jobbing hand would often be expected to design the layout of a handbill or an advertisement. It was, paradoxically, the newspaper hands, the best paid branch of the trade, who found themselves most fully confined to straight typesetting work and who therefore depended most heavily on their speed and dexterity as such.[8]

Beyond their technical indispensability, compositors buttressed their position in the division of labour through long traditions of workplace organisation and wider trade union structures. Drawing on trade customs, union rule-books, and scales of prices negotiated with employers (the famous London Scale of Prices was unilaterally enforced after the dissolution of the Master Printers' Association in 1866), the three regional typographical unions sought not only to fix the minimum wages and hours at which their members worked, but also to regulate working conditions in printing offices so as to ensure the long-term reproduction of the compositors' position in the division of labour. The unions therefore attempted to prevent the employment of non-apprenticed workers on any portion of compositors' work, to maintain a fixed ratio of apprentices to journeymen, to limit overtime and casual labour, to regulate the operation of piecework systems, and even on occasion to fix the maximum output of their members.[9]

Depending on their market position and on the nature of their product, printing employers sought to undermine this framework of regulation in a variety of ways. For newspaper proprietors enjoying rising circulations and advertising revenues, the problem was above all one of obtaining more rapid production, a consideration which applied to the weeklies and other periodicals with diminishing intensity depending on their publication schedule. Book firms, on the other hand, were less constrained by temporal pressures, but faced increasing domestic competition, often from provincial plants unfettered by union wage rates and work rules, which depressed profit margins and led them to place a higher priority on lower labour costs. Finally, for the small jobbing shops which made up a large proportion of the trade, urgent orders and low profit margins combined to create a double pressure on wages and work intensity which at its worst approached conditions in the sweated trades.[10]

Among the most important methods used by employers to speed up work and cut labour costs in the composing room was the manipulation of systems of payment and supervision. London had traditionally been a piecework centre, with composition paid at so much per 1000 ens according to the provisions of the London Scale; the provinces were dominated by time or 'establishment' work, with the 'stab' hand receiving a fixed minimum wage for a fixed working week. From the middle of the nineteenth century, however, there was a certain convergence, as 'stab' work penetrated London and piecework the provinces, although certain provincial unions strongholds remained 'stab' towns. it was not primarily the spread of new payment systems as such which was important, but rather the detailed features of their operation. The value of either system to worker or employer depended on how much work was expected for how much effort, as well as on how the burdens of fluctuations in demand and interruptions in production were distributed. Thus different payment systems proved more suitable to different types of work: in general, daily newspapers, where compositors were well organised, the pace of work intense, and employers able to pay, were worked on the piece, while the more varied and less remunerative work in book, jobbing and periodical offices was more often done on 'stab'.

It was therefore the growth of the 'dual system' or 'piece-stab' rather than the simple extension of piecework to 'stab' districts or vice versa which constituted the main thrust of employers' encroachments on craft regulation. This system, whereby 'stab' and piece hands worked side-by-side in the same office was disadvantageous to compositors in a number of respects. The distribution of copy was often manipulated so that the potentially well-remunerated jobs were given to the 'stab' hands and apprentices, while the piece hands were saddled with the more difficult and time-consuming work. Similarly, while the 'stab' hand was normally supplied with a steady supply of copy, the piece hand might spend a large portion of his time 'on the slate', idle and unpaid, a system often used by newspapers to keep compositors around at no cost to themselves in case of changes to late editions.[11] This practice could severely depress compositors' earnings, tending towards casualisation in extreme cases: in Edinburgh, where compositors were particularly badly organised, a union survey taken in 1890 showed that 26 per cent of its sample were earning less than 20 shillings weekly, while only 30 per cent were bringing home the full 'stab' rate of 30 shillings.[12]

'Piece-stab' likewise intensified the pace of work for all compositors, as piece hands strained to produce the maximum when copy was available, while employers expected their 'stab' hands to set more than the piece value of their wages if they hoped to retain their more secure position. Here changing patterns of supervision also played an important role. In many printing offices elected 'clickers', responsible for distributing copy and computing piecework earnings, gave way during the second half of the century to foremen appointed by management, and the more intensive supervision inaugurated by the latter went hand-in-hand with 'piece-stab' to create the system of high-pressure production decried by many compositors.[13]

Closely bound up with 'piece-stab' as a method of cheapening and intensifying compositors' labour was what one commentator termed its 'most pitiable, most deplorable and most degrading elder son', casual labour.[14] While the cyclical movement of printing output closely followed the business cycle in the economy as a whole, the industry was marked by its own specific seasonal cycle, which revolved around the movement of demand for advertisements and publicity, the London season and the parliamentary time-table. On daily papers, the unpredictable arrival of copy also produced sudden changes in the size and layout of editions which often necessitated the hiring of extra hands at the last minute.[15]

Musson estimates that in good years the proportion of provincial unionists in casual employment ranged from one-eighth to one-fifth, rising as high as one-third in slumps; a survey conducted by the union itself in 1892, an average year, showed a wide geographical range from 3.6 per cent in Oxford to 31 per cent in Manchester. It was where casual labour was combined with 'piece-stab' that the full depressive impact on wages of the latter made itself felt. Thus the 1906 Wages Census found that 28.8 per cent of all compositors and 33.7 per cent of piece hands were employed for more or less than the standard working week (some clearly working overtime); 60 per cent of the latter were earning less than 25 shillings per week, and 84.9 per cent of piece hands earning less than 25 shillings were working less than a full week.[16]

Systematic overtime furnished an additional source of underemployment and work intensification. The erratic and often urgent demand for printed matter combined with employers' efforts to maintain low manning levels to escalate pressures on the men employed for the rapid production of copy, leading to frequent and onerous overtime requirements. Thus a survey conducted by the London Society of Compositors (LSC) in 1892 found that more than half its members worked 'systematic overtime', and the provincial unions registered similar complaints.[17]

Given the impact of 'piece-stab', casual labour and systematic overtime on compositors' earnings and working conditions, the regulation of these practices figured prominently among the objects of the typographical unions during the second half of the century. Thus the unions consistently raised demands for the fair distribution of copy between 'stab' and piece hands and sought at various points to obtain 'standing time' payments for piece hands. Similarly, the unions refused to sanction any form of 'task work' — a minimum output required from 'stab' hands — arguing that high output should be paid for through piece rates, and opposed bonus payments as levers for work intensification. At the same time, the unions sought to maximise employment and limit the trend to casualisation by regulating the activities both of its own members and of the employers. Thus 'smooting' and 'twicing' (holding two jobs and moving from case to press, respectively) were prohibited, while union members were also discouraged from the employment of casual hands themselves, although London weekly newspapers were often farmed out to compositors responsible for hiring the staff. The unions likewise fought for minimum guaranteed earnings for casual labour and aimed to contain overtime through punitive rates and in some cases outright limits.[18]

Where compositors were well organised, as particularly on daily newspapers, these regulations prevailed to a greater or lesser extent, but in most cases they remained of limited effectiveness, as can be seen from the continued resurgence of similar demands. One source of the unions' difficulties lay in the persistent divisions among their members on the question of piecework, particularly outside London. While many provincial compositors attacked piecework on a combination of economic and moral grounds, news hands defended the practice as essential to ensure that the high levels of effort demanded by their work be rewarded by high earnings, and it therefore proved impossible to mount a unified campaign for its abolition.[19]

Perhaps the most crucial issue for the long-term effectiveness of craft regulation was control over apprenticeship. No subject was discussed more often at union delegate meetings or in union publications, and compositors were unanimous in placing the multiplication of boy labour at the centre of their difficulties in the labour market. Already in 1850 an authoritative estimate put the number of apprentices in the UK at 6000:8500 journeymen, though the situation was somewhat better in London where the figures were 1500:3000 respectively. It was not merely the quantity of apprentices but also their quality which alarmed the journeymen. The decline of indentured apprenticeships, and still more their diminished technical content, meant that apprenticeship was becoming more a source of cheap labour than a means of imparting technical training. The multiplication of badly-trained apprentices threatened not only the long-term state of the labour market, but also the immediate bargaining position of the journeymen. Thus apprentices might be favoured in the distribution of copy, or might actually usurp journeymen's jobs. Particular offenders in this respect were 'turnovers', nominally apprentices who had switched masters, but generally serving as partially trained cheap labour.[20]

Each of the typographical unions, therefore, sought to regulate apprenticeship and in particular to restrict the ratio of apprentices to journeymen. The provincial Typographical Association (TA) and the Scottish Typographical Association (STA) maintained sliding-scales by which the ratio of apprentices to journeymen varied with the size of the office up to a fixed maximum; only the LSC left the matter to customary regulation until the traditional 1:3 ratio was codified in 1895. In a similar vein, each union made a seven-year apprenticeship a condition of membership and required turnovers employed in society offices to be reindentured to a new seven-years term.

As the estimate for 1850 suggests, however, the typographical unions experienced considerable difficulty during the mid-century in enforcing their apprenticeship rules. The key problem lay in the multiplicity of small printing offices in country towns beyond the reach of union control. While well-organised printing houses in the major centres might adhere to the official ratio, the influx of former apprentices from the provinces not only undermined union control of the labour market but encouraged the larger employers to evade the apprenticeship regulations themselves. As a result, apprenticeship regulation was in disarray by the mid-1880s, at least outside of London. The provincial unions' submissions to the Royal Commission on the Depression of Trade reported that in most towns the number of apprentices approached if not exceeded the number of journeymen, with Liverpool alone of the large towns maintaining effective limits. The LSC's informal restrictions appear to have been somewhat more successful, with a sample in 1880 showing a ratio of 1:4 improving to 1:4.5 in 1890, though other evidence suggests an abundant stock of boy labour even in London.[21]

Apprenticeship regulation figured as the most prominent cause of disputes between unions and employers in the provinces, most of which ended in victory for the latter. As a result, the provincial unions were forced to adopt a more flexible approach in practice, enforcing the regulations where they were strong and turning a blind eye to violations where they were not, while excluding local typographical societies entirely where the number of apprentices had got out of hand. While in the TA moves to develop a looser but more enforceable apprenticeship scale were blocked by the stronger branches, the late 1880s saw the STA and the LSC successfully tightening up their apprenticeship regulations as part of a more general refurbishing of union organisation and expansion of membership. Contrary to the Webbs' belief that the Device of Restriction of Numbers was dying out among compositors, therefore, regulation of apprenticeship seems to have been reviving in the typographical unions from the mid-1880s onwards. It continued to be possible for boys to pick up the trade and even for non-apprenticed men to get work in many of the printing centres, especially in the TA's territory, but the STA was revitalising its restrictions and the LSC moving towards formal codification of its own limits.[22]

Compositors' struggles to control entry to the trade were not, of course, confined to apprenticeship regulation. Printing craftsmen's claims to the control of particular jobs necessarily implied the exclusion of other groups of workers. In contrast to engineering, where a previous wave of technical change had ensconced within the workshops a class of semi-skilled workers who could find themselves in conflict with craftsmen over access to particular jobs, the static division of labour in the composing room meant that craftsmen's competitors were drawn from outside the printing offices themselves, from the ranks of non-apprenticed men and, in several areas, of women as well.

The level of unionisation therefore provides a reasonably reliable guide to the effective coverage of craft regulation in the industry. The late 1880s and early 1890s saw a general movement in each of the unions towards the renovation of their strained frameworks of regulation through improved organising methods and increased local militancy. Full-time organisers were appointed in certain unions and in many areas full-time organising drives were launched. Alongside the tightening of restrictions on apprenticeship, the result was a rapid expansion of union membership: between 1880 and 1893 all three unions doubled in size, the LSC from 5100 to 10,151; the TA from 5699 to 12,736; and the STA from 1504 to 3004. By 1890, the LSC appears to have unionised some two-thirds of all London compositors, and only three of twenty daily newspapers were run on a non-union basis. In those towns where it maintained a branch, the TA claimed an 80 per cent membership in 1892, but given the sketchiness of its organisation in the south and south-west, Musson's estimate of two-thirds appears more plausible, perhaps even over-generous, as a global figure. In Glasgow the STA claimed a density of 95 per cent, but in Edinburgh its 800 members were matched by 400 male non-unionists and between 200 and 800 women, according to various estimates.[23]

Apart from non-unionised men, the main threat to the compositors' control of entry to the trade came from women. Thus the Census records an increase in the number of female printers (most of whom were compositors) from

700 to 4500 or 5.2 per cent of total employment. Apart from Edinburgh, the main areas in which women were employed were the country towns of England and Scotland; even in London there were probably 200 female compositors by 1890. In Edinburgh, where the system was most fully developed, women were introduced in 1872 after the masters' victory in a strike for the 51-hour week. The women were employed primarily in the large book houses, where they were usually confined to the more routine typesetting tasks, often on reprint or government work, while men performed the heavier tasks of making up and imposing, as well as some of the more skilled and less remunerative portions of the work. In one large Edinburgh book firm for which detailed information is available, the women served a three-year apprenticeship (elsewhere one year appears to have been more common) and their earnings averaged between 16 and 18 shillings, compared to the union rate of 32 shillings. The advantage to the employer lay not merely in the girls' lower basic rates, their employment allowed the masters to evade the onerous 'extra' payments and work rules of the union scales and, as one Edinburgh master testified, their presence on a wide scale weakened the union even in those shops where only men were employed.[24]

Wherever women compositors were widely employed, male printers were unanimous in regarding them as an important source of union weakness and of unemployment in their own ranks. Opposition to the use of women compositors therefore formed a cornerstone of union trade policy, though the unions were normally careful to proclaim their objections to *underpaid* female labour rather than to women *per se*. Women were to be excluded as a consequence of the general logic of craftsmen's efforts to restrict entry to the trade rather than because of sexual antipathy as such, though male compositors were prepared to countenance sexual discrimination to further their larger aims, and complaints occasionally surfaced about the inappropriateness of women doing 'men's work'. Thus while the unions expressed themselves willing to admit women earning the men's rate, and even backed a campaign to organise them in the extreme case of Edinburgh, their more usual response was exclusion wherever their local strength permitted.[25]

Surveying the printing industry on the eve of the introduction of the linotype, it is difficult to avoid the conclusion that the position of the hand compositor and his framework of regulation were increasingly embattled and in many respects deteriorating in the face of employers' pressures to cheapen and intensify his labour. Despite the pitfalls of nineteenth-century wages statistics, there is a certain amount of evidence that compositors' earnings were falling behind the rate of increase of the labour force as a whole, and of certain skilled trades in particular. Thus Bowley and Wood calculate that a weighted average of compositors' 'stab' rates rose by 21 per cent between 1860 and 1891, while Phelps Brown and Browne's index of money wages shows that overall wage rates rose by 43.5 per cent during the same period. It is more difficult to show that compositors were falling behind other skilled trades on this basis: Bowley's index numbers for fitters' and turners' wage rates show a similar increase of 22–3 per cent. The key here is the growth of 'piece-stab' which tended to depress earnings well below 'stab' rates for compositors in many districts. While compositors in regular employment on daily newspapers ranked among the best-paid manual workers in the country, these made up only some 10 per cent of the total even in London. As for the rest, compositors themselves were convinced that their earnings were falling behind those of other skilled trades, particularly the closely-related machine managers who did not have to contend with 'piece-stab', and this claim formed an essential component of their demands for wage advances after 1890.[26]

Even as the rise of 'piece-stab' and casual labour was undermining compositors' earnings and security of employment, the pressure of rapidly rising demand for printed matter on the traditional division of labour in the composing room was leading to an intensification of supervision and the pace of work equally subversive of their autonomy and craft status. By the late 1880s, high-pressure production had become an insistent theme of rank-and-file grievances, and commentators noted the difficulties of enforcing trade rules and customs even in union houses. Similarly, the decline in the technical content of apprenticeship appears to have resulted in a parallel deterioration in the quality of apprentices themselves. While this argument cannot be proved without more representative information on the changing social background of printing apprentices, it is supported by such fragmentary information as is available and serves as an index of popular perceptions of the compositors' status and prospects.[27]

Against this evidence of decline must be set the undoubted success of the compositors' unions in defending and even pushing back the frontier of control in certain areas, largely as a result of improved organisation from the mid-1880s onwards. Thus the unions were able to expand their membership dramatically and to extend their influence to new districts. Equally importantly, they were able to tighten their control over apprenticeship and to keep out female labour in most areas. This reinforcement of the compositors' exclusive position, itself a response to their deteriorating earnings and control over work organisation, would stand them in good stead during their struggles to capture new machinery during the following decade.

III

Much more so than in printing, the position of skilled workers in engineering approached the stereotype of mid-Victorian stability and prosperity during the two decades after 1850. As the rapid expansion and technical change of the 1830s and 1840s gave way to a period of extensive growth oriented towards export markets, a new division of labour solidified, within which specialised skilled workers occupied a central place. In this context skilled workers and their unions were able to re-establish on a customary basis many of the principles of craft regulation which had been overturned during the employers' offensive of the late 1840s and early 1850s.

The division of labour which prevailed in most engineering workshops during the second half of the nineteenth century was itself the product of a prior wave of technical change and industrial conflict during the 1840s and 1850s. As a result of their victory in the 1852 lock-out, engineering employers formally obtained the right to place labourers on machines, to employ non-union labour, and to impose systematic overtime; in many cases they also forced their employees to sign the 'document' renouncing trade union membership. The exceptional bitterness of this confrontation lay in the shared belief of skilled workers and employers, itself shaped by the preceding period of rapid technical and economic change, that the development of the division of labour might result in the eventual elimination of skilled craftsmen as a significant component of the engineering labour force.[28]

In the event, however, this belief, which has been echoed by historians such as John Foster,[29] proved incorrect for a combination of economic and technical reasons. The machines which were involved in the first wave of mechanisation in engineering — the slide-rest lathe, the planer, the slotter, the shaper and the driller — were capable of routinising only a part of the engineering craftsman's skills. Thus a new tripartite division of labour emerged in which the more routine machine work was allotted to specialised labourers known as handymen, while the highly skilled work of designing models and prototypes became the province of the patternmakers; the core of the labour force was composed of two types of skilled workers with intermediate skills, known collectively as engineers: fitters and turners, responsible for precise assembling and sophisticated work on lathes and other machine tools, respectively. Both of the latter groups drew upon a modicum of abstract knowledge of engineering technology to perform their particular tasks: the turner to execute complex operations on the lathe and to adapt general purpose tools to specific tasks; and the fitter to make his skill with the file compensate for the inability of the other tools to produce fully interchangeable parts. Both might be called upon to read drawings, to design special tools or fixtures, to grind their own tools, and to set their own feeds and speeds.[30]

The technical foundations of skilled engineers' position in the division of labour were further enhanced by the course of economic development after the mid-century. The rate of growth of domestic demand slowed after 1850 and export markets, providing steady but less spectacular growth, became progressively more important to the industry as a whole. As Burgess has shown, the period from the 1850s to the 1880s was marked by a pattern of extensive growth, in which labour-using investment broadly preserved the ratio of capital to labour and so brought in its train a considerable expansion of the skilled labour force: membership of the Amalgamated Society of Engineers (ASE), to take one index, grew from 9737 in 1852 to 72,221 in 1891. The turn towards overseas markets and the development of new products seem to have encouraged increasing reliance on skilled workers to meet less standardised demand, though this must be set against the trend towards specialisation noted by many contemporaries, the result of the development of the division of labour, however extensive.[31]

The changed economic context facilitated the emergence of a new pattern of industrial relations. Skilled workers could accept the presence of labourers on certain simple machines without immediate fear for the future of their craft, while employers could in turn accept collective bargaining and even concede many of craftsmen's demands on the organisation of work as well as on wages. Although the ASE had officially removed a fixed ratio of apprentices to journeymen and bans on 'illegal' men, piecework and systematic overtime from its rule-book after the employers' victory in 1852, the enforcement of these aims was entrusted to the districts, which by the early 1860s were enjoying considerable success in containing these practices. Thus an ASE survey taken in 1861 shows that the principle of the standard working day had become well established and systematic overtime contained in most districts. Similarly, the hated institution of subcontracted piecework known as 'piece-mastering' was tending to disappear, and piecework itself was confined to 10.5 per cent of ASE members concentrated in a few districts. The union was likewise able to insist on the payment of the district rate for work it considered to be skilled, and often to organise those who had entered the industry as handymen but who had been promoted to skilled positions.[32]

It would be a mistake to assume, however, that the stabilisation of the division of labour and the pattern of industrial relations after the mid-century ruled out serious conflicts between skilled workers and employers over the location of the frontier of control. Thus in 1866, for example, ASE members in a Manchester locomotive firm struck with Executive backing to demand the dismissal of a foreman whom they considered inadequately qualified. Similarly, the continuing fluctuations of employment caused by the intensity of the business cycle in engineering as a capital goods industry, together with new concerns for leisure time, gave rise to a broad and successful movement for the nine-hour day in 1871–2.[33]

By the 1870s and 1880s, moreover, signs of strain began to appear in the mid-century settlement, as the returns to a pattern of extensive growth diminished. There is evidence to suggest, for example, that labour costs were rising as a proportion of total costs as the increase in wage rates, particularly for skilled labour, began to outrun productivity growth in certain sectors. The onset of the Great Depression in the 1870s, bringing a shift in the terms of trade away from primary producers and hence a decline in overseas demand for engineering products, contributed to a squeeze on profits, as did the first stirrings of German and American competition in products such as agricultural machinery, locomotives and even steam engines.[34]

In the context of relatively stagnant demand, these pressures drove engineering employers to experiment not with new forms of capital-intensive investment, but rather with methods of cheapening and intensifying skilled labour within the existing division of labour, thereby reopening the whole range of issues whose conflictual importance had receded in the 1850s and 1860s. Foremost among these were the spread of piecework and the intensification of supervision as methods of boosting output and speeding up the pace of work without commensurate increases in labour costs. Employers sought to subvert collective bargaining and the standard rate by fixing piece prices unilaterally, and to capture the principal benefits from increased output by

cutting rates whenever workers exceeded an informal ceiling
on earnings. The pace of work might be speeded up by
setting the rates with reference to the faster workers, with
those unable to keep up weeded out. In the face of the
growth of piecework, the ASE reinserted a ban on its
extension into its rule-book in 1874, but major strikes on the
issue were defeated in 1876 and 1889. Thus ASE surveys
showed that 12.9 per cent of its members were paid by the
piece in 1876 and nearly 17 per cent in 1891, with the
proportion approaching 50 per cent in the West Midlands,
and higher in particular towns.[35]

'With the steady increase in the price of skilled labour,'
observed one employer in 1886, 'there is a constant conflict
going on, a perfectly peaceful one, between employers and
skilled labour . . . shown in endeavours to substitute skilled
labour by machinery.'[36] It was not on the whole the
introduction of new types of automatic and semi-automatic
machinery which were at issue, since few of these were sold
in Britain before the mid-1890s. Rather, employers tended
to make greater use of the possibilities already available
within the existing division of labour, occasionally promoting handymen onto lathes from planing machines, but more
often multiplying the numbers of simpler machine tools and
with them the burgeoning class of machinists, as well as
introducing 'fitters' assistants'. Thus in 1892 machinists were
70 per cent as numerous as turners in one large marine
engineering works, while fitters' assistants outnumbered
fitters and machinemen, and boys outnumbered turners in
another. Similarly, in the railway works of Crewe and Derby
and the cycle shops of the West Midlands, the 'team system'
had become prevalent, whereby a leading hand paced a gang
composed of several skilled men and a large number of boys,
often on piecework. Only in the smaller, less specialised
workshops (which still comprised a majority of the industry)
did craftsmen still hold unchallenged sway, as in districts like
the Potteries and the ports of the Bristol Channel.[37]

As in printing, the key to craft regulation lay in the
effective control of apprenticeship. In engineering, this
period saw an accelerated subversion of apprenticeship into
a form of cheap labour as the proportion of apprentices to

journeymen rose sharply and the technical content of their training declined. Though the ASE had rescinded its apprenticeship rule in 1852, the union sought none the less to enforce an informal ratio of 1:4. But a major strike in defence of this ratio in Sunderland was defeated in the early 1880s, and surveys conducted by the Webbs and the ASE in the early 1890s showed that only in a few districts did the apprentice ratio approach 1:4. In the small ports of South Wales and for a brief period in Hartlepool the ASE won an agreed ratio from the employers; elsewhere apprentices and boys proliferated, reaching 6:1 in some Scottish firms.[38]

Not surprisingly, such trends were most evident in those sectors where a significant degree of repetition production could be found, such as textile machinery, railway engineering, cycles, and especially armaments. The giant Newcastle-based firm of Sir W.G. Armstrong, Mitchell & Co. provides perhaps the clearest example. In 1890 the firm employed some 15,000 men in its naval shipyard, ordnance and engine works. While its battleships and large guns remained custom-built products requiring highly skilled labour, large orders for shells and machine-guns offered considerable scope for repetition production. Thus the firm made extensive use of machinists, and its managing director wrote to the Webbs, 'I take as many apprentices as I dare'; and an ASE survey found in 1894 that in five shops in the ordnance works there were 365 apprentices to 482 ASE members and 103 adult non-unionists.[39] More original were the firm's methods of labour discipline and supervision. By 1890 Armstrongs had already developed an elaborate system of time-keeping, using individual time-boards to monitor attendance and movements from day to piece rates. In shops engaged in repetition work, piece payment coupled with frequent rate cutting prevailed. But where the nature of the work precluded piece rates, as on large guns, or where the men were thought to be limiting output to prevent rate cuts, the firm introduced a special class of supervisors, known by the late 1890s as 'feed and speed' men, whose sole duty, according to one pro-management spokesman, was 'to keep moving through the shops in order to see that each machine is being kept at its proper speed and is producing the amount of work which it is known to be capable of turning out.'[40]

Even the most purely economic of the union's achievements — the standard working day and the standard rate — were threatened by the employers' assault on craft regulation. Between 1880 and 1890, unemployment in the metal industries (engineering, shipbuilding and iron and steel) ran some 26 per cent above the average for all industries. In this context, systematic overtime once again became endemic in the industry, as employers sought to take the fullest possible advantage of trade upturns during the overall depression, and to intensify the pressure on their workforces by keeping manning levels low. The cumulative force of these various tactics was to reduce the effective hold of district rates over earnings. Thus the 1886 Wages Census reported that, outside of London and Manchester, not only did average earnings for fitters and turners fall several shillings below the district rate, but some 15–30 per cent of the total were earning either 10 per cent above or below (in this context presumably below for most) average earnings.[41]

A final measure of the effectiveness of craft regulation in engineering lies in the extent of unionisation. Union density varied considerably by district and by sector. By all accounts, the ASE was strongest on the north-east coast, where estimates of membership ranged from 75–90 per cent of those eligible, though even there some plants were markedly better organised than others: thus Palmers shipbuilding and engineering works in Jarrow was 100 per cent organised in 1892, but Armstrongs only 60 per cent. Other ASE strongholds included marine centres such as Barrow, Birkenhead, Hull and the ports of the Bristol Channel, Lancashire, and the West of Scotland. In general, the ASE retained a reasonable hold on the older engineering centres, except for the railway towns. But in the newer centres of the industry — Coventry, Birmingham and London — the union had made little headway, and these districts evaded its control almost completely. In the industry as a whole, half of the 180,000 fitters and turners recorded by the 1891 Census remained outside the ASE, while the much larger numbers of labourers and handymen, some of whom had begun to organise their own unions, added to the pool of potential blacklegs.[42]

An additional source of weakness for the ASE lay in its persistent conflicts with related metalworking craft societies with whom they competed in some cases for members and in others for control of particular classes of work. Thus in addition to the Steam Engine Makers' Society (SEMS), which had held aloof from the ASE at the- time of its formation in 1851 and retained some 6000 members into the 1890s, other more specialised craft societies such as the various blacksmiths' societies and the United Patternmakers' Association (UPA) recruited groups eligible for membership in the ASE. The latter, formed in part as a breakaway from the ASE in 1872, maintained a special 'poaching clause' to attract the highly-skilled patternmakers away from the larger ASE which had difficulty in taking account of their special interests and strategic bargaining position. In retaliation, the ASE sought to extinguish the UPA by supplying blacklegs to take the place of its members, as most notoriously in the case of a strike in Belfast in 1892. Not surprisingly, the patternmakers often held aloof from ASE-led trade movements, and organised with other small societies a 'Federal Board' on the north-east coast in 1888 to avoid being drawn into disputes without consultation. Equally serious were the demarcation disputes which abounded with less closely-related trades, particularly in the shipyards. There the ASE's efforts to expand the employment opportunities of its members resulted in a wave of disputes with the other outfitting trades, particularly during the boom years of the late 1880s and early 1890s, of which the most virulent and destructive were those with the plumbers on the north-east coast in 1890–2. Similarly, while demarcation disputes with the boilermakers had faded away with the latters' consolidation of control over the main hull construction operations in iron and steel shipbuilding, the resulting heritage of rivalry and mutual mistrust, coupled with conflicts over collective bargaining tactics, caused the ASE to hold aloof from the Federation of Engineering and Shipbuilding Trades formed by the latter in 1890, thus reinforcing its isolation.[43]

By 1890, in engineering as in printing, employers' efforts to cheapen and intensify skilled labour had led to a

significant deterioration both of the effectiveness of craft
regulation and of the economic position of skilled workers,
without any major changes in the technical basis of the
division of labour. To be sure, the ASE remained one of the
largest and wealthiest unions in Britain, and like the
typographical unions it took advantage of the 1889–92 boom
to launch a counter-offensive based on local militancy and
organisational reform which enjoyed some successes in
restoring wage cuts, containing overtime and expanding
membership. In contrast to its counterpart in printing,
however, this movement made little progress towards the
re-establishment of effective apprenticeship regulation.[44]
Skilled engineers' lines of demarcation were complex and
difficult to police, an ambiguity which led them into conflict
with most of the related skilled trades with whom they
competed for jobs and members. Moreover, in contrast to
printing craftsmen's unchallenged control over the composing room, a previous wave of technical change had ensconced within the engineering workshops a class of less
skilled men whose promotion prospects were blocked by the
ASE's exclusive practices and who would become a real
threat once further technical change called into question the
existing division of labour. Thus the onset of a renewed
wave of technical and organisational change in the 1890s
would find skilled engineers and their union badly placed to
maintain their position in the next phase of the division of
labour and vulnerable to a major employers' offensive.

IV

While printing employers during the second half of the
century concentrated on cheapening and intensifying skilled
labour in their efforts to overcome the bottleneck to the
rapid expansion of output represented by traditional typesetting methods, they were acutely aware that a true solution
to the problem of rapid, cheap and efficient composition
would require mechanisation. Accordingly, newspaper
proprietors experimented repeatedly with various composing machines from the 1820s onwards. While rudimentary

composing machines such as the Hattersley and the Thorne, based on the principle of a keyboard which dispensed individual moveable types into the compositor's stick, were introduced in small numbers onto provincial papers (and the non-union *Times* in London) by proprietors anxious to speed up composing room production, these did not prove adequate to the demands of daily newspaper work. Beyond their technical defects, these machines failed to reduce overall labour costs, requiring cheap labour in redistributing the used types to the magazine to break even, and adding costs where regular compositors were employed on all tasks, as at the *Daily News* in London. Stressing the intellectual content and variety of the hand compositor's work, a leading trade journalist outlined in 1890 the essential features for the successful composing machine of the future: speed, accuracy, simplicity and automatic justification and distribution. While initially the linotype could not meet these requirements, the models in operation by the early 1890s did.[45]

Invented in New York in 1886 and rapidly diffused throughout American newspapers, the linotype, unlike its predecessors, did not use moveable types of the kind employed by the hand compositor; instead, it cast each line as a whole from hot metal. When the compositor manipulated the keyboard, a series of matrices were released from the magazine which cast the letters in slugs of molten lead. The matrices automatically returned to their place in the magazine, while the used slugs were simply returned to the melting pot, solving the problem of distribution. The matrix for each letter included the appropriate spacing, eliminating the need for internal justification, though the operator remained responsible for hyphenating and justifying the columns. At the same time, the linotype proved capable of setting an average of 6000 ens per hour, as much as four or five hand compositors.[46] As a result, these machines quickly swept the field of their rivals: according to a union survey, by 1895 there were over 250 linotypes operating in the provinces to 33 Hattersleys and 14 Thornes. By February 1896, according to the Linotype Co.'s own figures, 157 British printing establishments were using a total of 743 linotypes.[47]

While many printing employers later came to believe that the linotype really required no more skill to operate than did a typewriter, the failure of previous attempts at mechanical composition disposed them to believe that fully-trained compositors were required as machine operators, a view encouraged by the Linotype Co. itself, which initially hoped to win union cooperation in the introduction of its product.[48] It is at the same time significant that the new machines were first introduced in newspaper offices whose proprietors were sharply divided by commercial rivalries, notoriously vulnerable to union pressure because of the perishable character of their product, and inclined to prefer compromise to confrontation by their buoyant market position. Wherever linotypes were introduced, therefore, these new composing machines like their predecessors were captured by union compositors and subsumed under their framework of craft regulation. By the time printing employers had realised the productive potential of the linotypes and had revitalised their collective organisations, union control of the machines was already securely entrenched, though employer pressure was able to secure the relaxation of union regulation to a greater or lesser extent depending on the region.

The basic strategy underlying the response of skilled compositors and their unions to the rapid diffusion of the linotype was relatively uniform across the country. In each of the three main regions, the unions sought initially to maintain the *status quo ante* as far as possible without actually opposing the introduction of the machines. In their view, the existing framework of regulation should be strictly applied to the machines so that neither hand nor machine compositor should be disadvantaged, while the men should also receive a fair share of any benefits from increased productivity. Thus in local negotiations over the introduction of composing machines the unions demanded that all machines should be operated by fully-trained compositors or apprentices in the last years of their terms, with preference given to existing employees in each office; that working hours be cut and wage rates advanced; that trainees receive the full 'stab' rate for an extended period while other compositors were forbidden to learn in their spare time. Any form of incentive to increased output other than high piece rates, such as 'task work' or bonuses, was likewise prohibited.

This policy of strict regulation soon brought the unions into acute conflict with the employers, first on an individual basis and then, in London and the English provinces, with newly organised employers' associations. In each case, this collective employer pressure forced the unions to relax their framework of regulation in order to preserve overall control of the machines. The precise character of the settlement in each region depended on prior traditions of collective bargaining and on the forces shaping the balance of power between skilled workers and their employers, notably the degree to which employers were able to act collectively on the one hand, and the degree to which unions were willing and able to mobilise their members on the other.

In the English provinces, where the linotypes were first introduced in large numbers, the TA Executive sought to regulate their operation by formulating, in consultation with rank-and-file delegates, a set of guidelines for local bargaining. The restrictive character of these regulations — particularly their provisions for tuition payment, high piece rates and barriers to increased output — soon provoked employer opposition, leading to victimisations of union activists and short stoppages in a number of towns. Employers did not long remain content to pursue their resistance individually, and a series of informal conferences culminated in the formation of the Linotype Users' Association in 1894, which quickly came to see a national agreement on the machines as the solution to its problems with the unions.[49]

While the TA was initially unwilling to abandon its tradition of local bargaining which allowed it to concentrate its resources on isolated employers, a number of developments began to weaken its resolve. Disagreements between the LSC and the Linotype Co. over tuition restrictions had provoked the latter to open a training school in London for non-union machine operators, and a second school was planned for Manchester. Chastened by the emergence of this first real threat to union control of the machines, by its inability to enforce the 40 shillings minimum 'stab' rate laid down by the delegate meeting, and by rising expenditures on dispute benefits, the TA Executive, with the reluctant support of the rank-and-file-based Representative Council (RC), led a strategic retreat from the tight restrictions previously established over tuition payment and the 'dual system' (case and machine hands working alongside each other), as well as lowering 'stab' rates. After several years of negotiations, the TA Executive, faced with sustained employer pressure, went on in 1898 to accept a national agreement which fixed 'stab' rates at 12½ per cent above existing case rates and working hours at 48 for day- and 42 for night-work (but without a higher rate for overtime). Most importantly, however, all operators were to be TA members, with preference for those already employed in each office. Apprentices could work the machines after three years at a ratio of no higher than 1:3 journeymen; and operators were guaranteed three months' 'stab' wages before being placed on piece rates.[50]

The 1898 settlement was widely criticised by rank-and-file compositors fearful that the relaxation of strict regulation would exacerbate the displacement of hand workers (running at some 20 per cent during the 1894–6 recession, and concentrated among older men) and irate at the Executive's unilateral abrogation of policies formulated by the delegate meeting. But given the relative weakness of the union's position, and the absence of any coherent alternative poliy, this opposition quickly fizzled out. A censure move by the RC was overturned by a membership ballot and Executive hegemony in this traditionally oligarchic union was re-established by the dissolution of the former body soon after.[51]

The progress towards collective agreements on composing machines in London was similar in many respects to that in the provinces. Two fundamental differences relative to the provinces tilted the balance of forces in favour of the compositors in the metropolis: the deeper fragmentation of London newspaper proprietors by competitive pressures compared to their provincial counterparts, and the more extensive rank-and-file participation in union decision-making in London. These factors, coupled with the superior effectiveness of union organisation which was in large measure their consequence, ensured that the settlement eventually reached over the introduction of composing machines was substantially more favourable to skilled compositors in London than in the provinces.

While collective bargaining in London printing dated back to the late eighteenth century, the London Master Printers' Association (MPA) had been dissolved in 1866 and only reorganised in 1890, principally as a result of the large book and jobbing firms' opposition to the LSC's demands for a major revision of its Scale of Prices. It was at the behest of the newspaper proprietors newly affiliated to this body that the LSC reluctantly accepted in June 1894, soon after the first linotypes had been introduced in London, a provisional composing machine scale fixed to run until the end of 1895. As in the provinces, this agreement specified that all operators should be LSC members with preference given to hand compositors in the affected offices, but it also recognised substantially higher piece rates than those demanded by the TA and offered case hands working alongside machine operators the added protection of a simultaneous 'lift' and 'cut' (start and finish of work) to ensure fair distribution of copy to each group. It is evident from these terms, which were exceptionally favourable to the union (indeed, the LSC later argued that the employers had been 'caught napping') that newspaper proprietors did not at first recognise the full potential of the linotype nor did they intend to use it as a wedge to escape from craft regulation, which instead immediately engulfed them.[52]

Dissatisfaction with the hastily agreed 1894 machine scale spread rapidly among employers with the accelerated diffusion of the machines themselves. Here the Linotype Co., which regarded the 1894 agreement as a major brake on the sales of its product, played an important role, rallying London newspaper proprietors to demand a revision of its provisions and as we have seen, establishing a training school to create a pool of non-union labour. As employer resentment mounted over high piece prices and low output, a group of London newspaper proprietors under the leadership of the *Daily Telegraph* served notice on the LSC that the 1894 scale would not be renewed. The LSC Executive, experiencing some difficulty in retaining control of the machines as a result of the Linotype Co.'s training school, was quite prepared to negotiate, and a provisional agreement on a revised composing machine scale was reached in December 1895. The key changes involved a small reduction in piece prices, the reduction of the paid tuition period to four weeks, and the abandonment of simultaneous lift and cut for case and machine hands, though a guaranteed minimum wage was introduced to protect the former. At the same time, the revised scale formally recognised for the first time a 1:3 apprentice ratio in weekly news and book offices (they had long been banned entirely from daily papers).[53]

As in the TA, this relaxation of strict regulation of the machines ran into serious opposition from rank-and-file compositors concerned about the displacement of hand-workers. But in the geographically compact and relatively democratic LSC, the membership played a much larger role in union decision-making. Thus policy was formulated by quarterly delegate meetings, general meetings were regularly convened and well attended, and membership ballots not infrequently reversed Executive decisions.[54] Hence, when the trough of the trade cycle coincided with the peak of labour displacement at the hands of the machines in 1894–5, LSC activists, led by the Unemployed Chapel, were able to press with considerable success for more militant trade policies and tighter enforcement of union rules as a remedy for unemployment. Early in 1894, the membership approved not only a formalisation of the apprentice ratio, but also its reduction from 1:3 to 1:6; and later that year a general meeting affirmed the urgency of a movement to obtain the 48-hour week. Similarly, a meeting of the news department, which had led an independent existence until 1854, announced its opposition to any dilution of the principle of simultaneous lift and cut, while a ballot of the membership endorsed a proposal to contain unemployment and the intensification of work by prohibiting any member from producing 'more than has been agreed to by the chapel as . . . a fair day's work'. When the Executive submitted the revised machine scale for ratification to a special delegate meeting, it was rejected, and a general meeting in January 1896 endorsed this decision, voting a strike levy by a large majority.[55]

While the employers had threatened a massive closure of their offices to LSC members in the event of a rejection of the revised scale, in the event the predicted general collision failed to materialise. While one London daily did attempt to impose the new scale unilaterally, backed by blacklegs supplied by the Linotype Co., the rest of the newspaper proprietors caved in when faced with the evident militancy and determination of the workforce. Even during the original negotiations, a group of visiting American newspapermen had noted the evident 'want of unity among London newspaper owners', and the spectre of the large, irrecoverable losses and potential threat to their volatile market shares entailed by a prolonged stoppage rapidly dissolved the employers' front. The striking profitability of the linotype, even under union restrictions, provides an additional explanation of the employers' moderation. Estimates of the savings on composing room costs effected by the machines ranged from 20–40 per cent despite the operators' high wages, and an authoritative study from the following decade when the machines' capacities had become better established, put the savings at 66 per cent over hand labour. The greater speed of linotype production also meant that papers could increase their circulation and so gain more than their actual saving on labour costs. In this context, the London employers found themselves prepared to accept a redrafted machine scale which preserved the protection of simultaneous lift and cut for handworkers in return for a minor reduction in piece rates (itself only one-third of their original demand).[56]

While in London and the English provinces the introduction of the linotype led to collective agreements with employers which ratified such crucial components of craft regulation as the restriction of machine work to fully-trained union compositors and the recognition of a fixed ratio of apprentices to journeymen, local bargaining remained the

rule in Scotland. Much more than the TA, the STA
remained a loose federation of autonomous local branches
enjoying widely different wage rates and working conditions. In Glasgow, where union organisation was strong and
conditions approximated to those in the major English
cities, after an abortive attempt by one newspaper owner to
impose low piece prices the local typographical society
obtained agreements from the employers which gave them
control of the machines on terms similar to those won by
their English counterparts. Elsewhere in Scotland, the
relative success of employers in cheapening and intensifying
hand labour through the use of women and 'piece-stab'
meant first, that the linotype was introduced more slowly
than across the border, and second, that once introduced
machinery often made possible an improvement in the
position of the compositor by alleviating the pressures on
employer profitability. In a number of Scottish towns, the
introduction of composing machines in the late 1890s
appears to have resulted in a decline in the demand for
casual and female labour, and in a reduction in the number
of apprentices.[57]

While the composing machine agreements of the late
1890s recognised union control of the linotype, they did not
represent a final settlement of the fate of craft regulation in
the industry, but rather provided the framework within
which skilled workers and employers could contest the shape
of the emerging division of labour. Though dissatisfaction
with the terms of the agreements abounded among both
union members and the employers, it was the latter who
made the initial running in pressing for revisions of the
linotype scales as their sense of the productive capacity of
the machines deepened and their collective organisation
improved. In the provinces the employers insisted that the
12½ per cent increase in 'stab' rates on the machines granted
in 1898 represented a maximum rather than a minimum, and
proceeded to cut wages in several towns, forcing compositors to make up lost earnings through increased overtime, so
forfeiting the advantages of the shorter working week. More
importantly, their dissatisfaction with output on the
machines led the proprietors to introduce indicators on

composing machines and bonus systems despite fierce union opposition. In 1903 the LUA won a machine piece scale from the TA Executive, though its unattractive terms led employers to abandon it in practice.[58] In London, where the employers were less united and the Executive more willing to risk strike action because of the militant stance of its members, the employers' offensive was less successful. Despite lock-outs on several papers, London newspaper owners were unable either to stem the course of wage drift by imposing a machine 'stab' scale or to break the hold of the union's restrictive work rules.[59]

The most serious challenge to craft regulation in the industry came, however, from the large book firms, whose ability to stockpile their relatively standardised product allowed them to contemplate a prolonged stoppage with greater equanimity, and whose declining profit margins and vulnerability to low wage provincial competition forced them to take a harder line on labour costs. While the large London book firms had led the resistance to general wage advances in 1890 and 1900, the sharpest clash arose over the introduction of the monotype in the years after 1900. A number of large book firms had introduced the linotype, but they found it less suitable for their special needs than did the newspapers, and it was not until the invention of the monotype, which employed reusable punched tapes to cast individual lead types, that book production became widely mechanised. Having learnt from the experience of the newspapers with the linotype, the book firms were determined to introduce the monotypes on more advantageous terms. Thus in London, the MPA rejected out of hand the LSC's proposals for a monotype scale, refusing any consideration of piecework and pressing for a low 'stab' rate.[60] Similarly, under the auspices of the recently formed Federation of Master Printers (FMP), the provincial employers won an agreement from the TA which stipulated lower piece rates and longer hours than the linotype scale and, more importantly, denied the union exclusive control over not only the less skilled casters but also over the keyboards.[61] Even more disturbingly, some employers in all three regions believed that the keyboards could best be operated by women, and introduced in their works an alternative division of labour on the Edinburgh model, with a small number of skilled men supervising a larger number of women on the more routine work. An attempt to introduce women operators at one London book firm during the favourable legal climate prevailing after Taff Vale was frustrated by a prompt strike in which the compositors were supported by the new unions of the less skilled affiliated to the London Printing and Kindred Trades Federation (PKTF) which had been formed at the end of the previous decade, and the LSC generally retained control over the keyboards, though no monotype scale was signed until 1923. In the provinces, too, the union retained overall control though female monotype operators persisted in certain offices.[62]

North of the border, however, the introduction of significant numbers of female monotype operators touched off a movement against underpaid female labour in Edinburgh which ultimately succeeded in removing this historic source of local union weakness. Following a movement initiated by the stronger branches of the union in 1904, compositors in a number of Scottish towns secured the abolition of low-paid female labour (principally on hand work) as mechanisation alleviated the most severe pressures on employer profit margins. The threat posed by female monotype operators to the already embattled Edinburgh compositor, coupled with expanded inter-union cooperation resulting from the reunification of the compositors and machinemen and the establishment of an effective Printing Trades Federation, prompted a demand for the abolition of underpaid female labour in 1909. After a successful strike the following year, in which the compositors were supported not only by the other printing unions but also, surprisingly, by the women themselves who hoped to improve their wages, the master printers conceded a ban on further recruitment of women, particularly on monotype keyboards, and a wage advance for the women. These terms soon became permanent, as the remaining women were organised into a special section of the union.[63]

In London and the English provinces this period also saw

major federated advance movements which contributed significantly to the consolidation of craft regulation. In contrast with engineering, where conflicts between craftsmen and the less skilled over promotion lines inhibited all grades movements, the greater success of skilled workers in excluding competitors from the composing room allowed them to support the formation of the new unions of less skilled printing workers formed in 1889. The intensification of industrial conflict during the 1890s led, after several failures, to the formation of Printing Trades Federations first in the provinces and then in London, and ultimately to the establishment in 1901 of the National Printing and Kindred Trades Federation (NPKTF) with an independent strike fund. The NPKTF quickly emerged as the principal focus for joint movements for the shorter working week in various provincial towns during the early years of the twentieth century, while in London the PKTF was instrumental first in winning employer recognition for the unions of the less skilled, and then in repulsing attempted lock-outs of LSC members over women on monotypes in 1905 and over the closed shop in 1906.[64]

With the revival of serious unemployment among compositors during the recession of 1901–10, rank-and-file-based pressures for a tightening-up of craft regulation converged with the growing sentiment in the printing trades as a whole to fuel a federated campaign for the 48-hour week. In London, compositors, elated by their victories over the employers in 1905–6, pressed forward to impose new restrictions on casual labour, overtime, bonuses and the remaining vestiges of 'piece-stab'. In the provinces, the upsurge of rank-and-file sentiment took the more defensive form of a revolt against the concessions on working practices granted during the preceding decade, as the 1908 delegate meeting unilaterally revised the TA's rule-book to prohibit indicators on composing machines and to impose strict limits on overtime. Since the 1890s the shorter working week had figured prominently among compositors' proposed solutions to persistent unemployment and work intensification, and ballots of the federated unions in 1908 showed a clear majority of the membership in favour of a joint movement to secure the 48-hour week.[65]

As the federated movement got underway, however, tensions between the London and provincial unions soon became apparent. The London unions were riding the crest of a wave of successful militant action and faced a divided body of employers, the newspaper proprietors having granted the 48-hour week in the 1890s and withdrawn from the MPA in the face of a threatened city-wide strike in 1906. The TA, by contrast, had been forced into a series of tactical retreats over the regulation of composing machines, and now found itself faced with an unprecedented national coalition of printing employers from all sectors, united by their opposition to the shorter working week and the new union rules, and by a desire for a relaxation of the apprenticeship scale.[66]

Ultimately, therefore, the London unions struck independently in January 1911 for a modified demand of a 50-hour week. Though the contest was ostensibly about hours, its intensity was dictated by a concern on both sides for the future of craft regulation. The men explained the urgency of the demand as a response to unemployment and work intensification related to mechanisation; while the employers justified their intransigence by the increased 'dictation' of the unions and their work rules, particularly regarding composing machines. The united front of the various London unions soon subdued a wide range of employers, including some of the larger firms, but a coalition of large book and small jobbing firms held out successfully on a non-union basis until the termination of the strike in July. This partial stalemate was unblocked the following year by the Treasury's decision to make the 50-hour week compulsory for government contractors on the basis of a survey showing that these hours were being worked by a majority of London printing workers. The hold-out firms were reunionised during the period of tight labour markets before and during the first world war, and this strike marked the end of effective resistance by London printing employers to the growing hold of craft regulation over the industry.[67]

The example of the London strike proved a powerful

force for compromise in the provinces, and a settlement was reached on the basis of 51 hours. The TA agreed to a limited extension of the apprenticeship scale, and accepted indicators subject to guarantees against abuses. Bonuses, task work and copy marking were prohibited, and a slightly relaxed overtime limit remained in force, while a further revolt at the 1913 delegate meeting won the replacement of the indicators by time-sheets. This settlement was capped by the creation of a joint conciliation board to which all disputes were to be referred before resort to strike or lock-out. The favourable character of this arrangement, subsequently accepted by the other typographical unions, can be seen from the fact that, in contrast to the disputes procedure imposed on the ASE by the engineering employers, the conditions to prevail during the procedure were those which had existed *before* any managerial initiative, thus representing no threat to craft regulation.[68]

By the eve of the first world war, the conflicts in Britain over the introduction of composing machines had clearly been resolved. In each of the major regions, hand compositors had gained effective control over the machines on a craft basis, though the terms of the settlement varied with the bargaining power and political style of the various typographical unions. Formal agreements with employers restricted linotype work to fully-trained union compositors, while union control of the monotype was first won *de facto* and then recognised *de jure*, at least in London and Scotland. This capture of the new machinery by hand compositors, in the context of improved union organisation, increases in the speed of production and reductions in labour costs resulting from mechanisation, brought in its train a marked increase in the effectiveness of craft regulation.

The increasing hold of craft regulation was most evident in the tightening control over the labour market exercised by the unions. Pockets of non-unionism were steadily eliminated, despite temporary set-backs such as the secession of some London houses in 1911, and in Scotland the advent of mechanisation had made possible the elimination of underpaid female labour. It was the steady revival of apprenticeship, already underway from the 1880s, which represented the most important strengthening of craft regulation. Whereas it had often been possible for mid-Victorian compositors to pick up the trade in small country printing offices, chapels were now stricter in demanding proof of new men's credentials, and in any case machine skills could only be acquired in the larger offices able to afford the initially expensive new techniques. The initial composing machine agreements made explicit provision for apprenticeship restriction, and the regional agreements signed at the end of the period incorporated modified versions of the unions' scales. By 1914, there had emerged a fair body of evidence showing that these ratios were being effectively enforced throughout the country. Surveys by the unions and the Board of Trade put the ratio at 1:4 in the English provinces, and 2:5 in Scotland; in London, where apprentices were banned from daily newspapers, apprenticeship regulation was universally acknowledged to be even more effective, and a survey in the mid-1920s put the ratio at 1:7.5.[69]

The consolidation of craft regulation likewise made it possible to reverse the deterioration in the economic position of the compositor which had been evident before 1890 as a result of the employers' need to cheapen and speed up hand labour. The introduction of composing machines removed the most urgent pressures of this kind and created a space for concessions on working conditions in the face of intensifying union demands. Thus the increased speed of the machines reduced the demand for overtime and casual labour, a development which coincided with a growing restrictiveness in union regulation of these practices aimed at soaking up labour displaced by mechanisation. With the growth of collective agreements some version of these rules was accepted by employers in each region, and they seem to have been fairly effective — Beveridge estimated that in 1908 the casual fringe among compositors had fallen to 15 per cent from closer to a third twenty years earlier. Similarly, despite the initial displacement of hand workers, the protection of craft rules providing for equal access to copy in due course reduced the abuses connected with 'piece-stab', so that complaints about casualisation and slating were increasingly confined to districts where mechanisation remained underdeveloped, such as Edinburgh. The rapid growth of the printing industry during this period meant that much of the hand labour displaced by composing machines could be absorbed in those composing room tasks which remained unmechanised, such as making up, imposing and display work, or by the increasing number of machine operating jobs themselves. With the elimination of these historic sources of weakness through the consolidation of craft regulation, compositors were able to reverse the trend for their earnings to fall behind those of other skilled trades, emerging as the best paid group in the labour force between the wars.[70]

V

In engineering as in printing, a series of interrelated developments in the 1890s were calling into question the existing structure both of the division of labour and of industrial relations. The gradual exhaustion of returns from an extensive development of the division of labour, the intensification of foreign competition, and new opportunities for mechanisation came together in the context of heightened conflict between employers and skilled workers over routine issues to precipitate a full-scale crisis in the existing pattern of industrial relations.

The stability of the pattern of investment, and with it the division of labour in engineering after 1850, had rested on two related economic conditions: the commanding position of British products in world markets and the weak demand in Britain for standardised, mass-produced engineering goods. Both of these conditions were eroded to a significant extent in the 1890s, with the increasing penetration of American and German products, first into Europe and the undeveloped countries' markets, and then into Britain itself, and with the diffusion of new American machine tools in the wake of the bicycle boom.

From the 1840s, American manufacturers operating behind steep tariff walls developed a new generation of automatic and semi-automatic machine tools to cater for a burgeoning demand for mass-produced consumer durables — small arms, agricultural machinery, typewriters and sewing machines — for which the American market provided greater support than did the British. While American methods — turret and capstan lathes, milling and grinding machines, precision gauges — made it possible in principle to routinise a much larger proportion of skilled labour in the industry, their impact on British engineering practice before 1890 was minimal. Neither the diverse character of most engineering demand nor its slow growth had encouraged British engineering employers to launch a major programme of capital-intensive investment.[71] Hence, as S.B. Saul has argued, it was not until the bicycle boom of the mid-1890s that a broad-based demand emerged in Britain for a product with standardised, interchangeable parts; and it was this demand which effected the diffusion of American machine tools, first within the cycle industry itself and then in the older sectors as well.[72]

Though the new machine tools were best suited to mass production, they were flexible enough to be used on less standardised work as well, since in some areas they extended the technical capacity of the engineering workshop in absolute terms. The intensification of foreign competition, together with the example of the cycle-makers and the falling prices of the new machine tools and the improved network of distribution in Britain, encouraged manufacturers in the older sectors of the industry to experiment with the new techniques. In most cases, this amounted to the piecemeal introduction of new equipment, perhaps occasioned by normal depreciation of older plant, rather than wholesale scrapping of existing machinery in the interest of a transformation of the division of labour. The process went furthest in those older sectors most involved in repetition work, such as armaments and textile engineering, but also touched marine engineering, locomotive building and even machine tool-making. Though the practical consequences of the new techniques remained limited in the late 1890s, the result was none the less to call into question the position of skilled craftsmen within the division of labour.[73]

This new wave of mechanisation inaugurated during the

mid-1890s coincided with a period of intensified conflict between skilled workers and their employers. With the revival of trade from the late 1880s, skilled engineers launched a major offensive to regain ground lost to the employers during the depression on a broad front of issues ranging from wages to the regulation of apprenticeship, machine manning, piecework and overtime, ultimately looking forward to the conquest of the eight-hour day.[74]

While disputes flared up in London and Lancashire, the movement was concentrated above all in the marine districts, where connection with the sharp cyclical fluctuations of the shipbuilding industry encouraged workers to take the fullest possible advantage of the leverage afforded by a boom. Thus ASE district committees in the Bristol Channel won limitations on overtime and the number of apprentices, while skilled engineers enforced a closed shop in the Barrow naval shipyards.[75] The real storm centre, however, was the north-east coast, which by the early 1890s had become the best organised district in the ASE. A newly-formed coordinating committee of ASE districts in the region quickly obtained an advance of wages and a small reduction in the working week; in 1890–1 it embarked on a campaign for the restriction of overtime which enjoyed some limited success despite a protracted lock-out. This cycle of industrial conflict culminated the following year in a series of acrimonious demarcation disputes, provoked both by the ASE's broader efforts to swallow up the smaller sectional societies and by the employers' attempts to manipulate the latter against the ASE.[76]

Following a series of organisational reforms in the ASE in 1892 and the revival of concern about unemployment with the downturn of trade after 1893, the focus of union militancy shifted to a national campaign for the eight-hour day. By 1894, the eight-hour day had been conceded in all government establishments, and major private employers had followed suit in a variety of engineering centres. Thus the union appeared on the verge of achieving a universal 48-hour week when the deepening of the recession brought the movement to a halt in 1895.[77]

It was this resurgence of craft militancy from the late

1880s which provided the impetus for the creation of a centralised national employers' federation in 1896. The Iron Trades Employers' Association (ITEA), the product of a previous attempt to form a national association of engineering employers after the nine-hours strikes of 1871–2, had foundered on the division between inland and marine producers, and despite the affiliation of the north-east coast firms in the mid 1880s, it remained a loose confederation of local associations without the power to undertake national action. Coordination among the local employers' associations on the north-east coast had developed out of a major strike in Sunderland over apprenticeship in 1883–5, and by the following decade employers throughout the region were operating a unified system of blacklists against union activists. In the course of the disputes with the ASE in 1890–2, the north-east employers had developed the tactic of turning isolated disputes into regional lock-outs, had established an elaborate mutual strike insurance scheme, and were seeking links with engineering employers in other districts.[78]

The final impetus towards national organisation came, however, from the success of coordinated action in the 1895–6 Clyde–Belfast dispute, in which Belfast employers were able to resist a demand for a large wage advance by means of a sympathetic lock-out on the Clyde.[79] This example provided the spark for the formation of the Engineering Employers' Federation (EEF), which at first drew in the employers' associations of the major marine centres alone. The intensified conflicts of the early 1890s over routine issues such as overtime, apprenticeship and demarcation, had prepared the ground for this breakthrough towards national organisation, but the belief that mechanisation might provide the opportunity for a significant transformation of the division of labour — which had been absent among printing employers — likewise stiffened the employers' resolve. The new organisation, whose Executive was endowed with centralised decision-making powers and authority to subsidise firms for strike losses, was explicitly aimed at restoring the hegemony of managerial prerogative and at providing a framework for collective action in response to demands raised by workers and unions over a range of issues from workplace organisation to wage bargaining.[80]

With the accelerated diffusion of the new machine tools after 1896, the problem of machine manning moved rapidly to the centre of the mounting tensions between the ASE and the newly-established EEF. From the outset, the ASE, like the typographical unions, realised that the capture of the new machine tools was essential if the skilled craftsmen's position in the division of labour was to be maintained; and a number of strikes against the promotion of handymen onto machines ensued. If anything, the ASE's position on the introduction of machinery was more moderate than that of the compositors, oscillating between the demand that the machines should be reserved for skilled engineers, and the more defensive claim that whoever worked them should be paid at the skilled rate. But by late 1896, the EEF had decided to mount a coordinated resistance to union machine-manning claims, taking their stand on the principle of the rights of property.[81] Thus a series of disputes in various districts over machine manning, 'interference' with foremen, restrictions on overtime, and payment for trial voyages to test new warships led the Federation to threaten a national lock-out in March 1897.[82]

While the more routine issues such as overtime and trip trials were quickly resolved through negotiation, the 'machine question' proved less tractable. In 1896 the EEF had been prepared to conceded control over certain machines to the ASE — principally sliding and screw-cutting lathes and large boring machines — in exchange for a free hand with the rest. But by the time of the machine conference in April 1897, this willingness to compromise had disappeared. The ASE Executive, aware of its vulnerable position, was for its part prepared to treat the machine question as essentially a problem of wages, and therefore proposed the establishment of local joint boards of workers and employers under Board of Trade arbitration to adjudicate which machines should be worked at the skilled rate. These proposals, which were opposed by the delegate meeting in June, would have proved extremely difficult to implement in practice, since the class of work rather than the type of machine was the key determinant of skill requirements. But in any case, the employers, conscious of their growing strength, had become committed to an unyielding defence of managerial prerogative, and the conference therefore broke up without reaching agreement, leaving the lock-out threat still outstanding.[83]

In the event, however, it was not the machine question itself which triggered the impending national lock-out, but rather the resumption of the movement for the eight-hour day outside the EEF's sphere of influence in London. A joint committee of engineering and shipbuilding trades, having secured the eight-hour day from a majority of large London firms, announced its intention to bring round the hold-outs with an overtime ban, to be followed by a strike in July. Faced with this ultimatum, the loosely organised London employers applied for membership in the EEF, and the Federation responded by threatening a national lock-out if the London strike notices were not withdrawn. The ASE, backed by several smaller unions, held its ground and the lock-out began in early July.[84]

Though the labour market was booming and the ASE's membership and financial resources had reached unprecedented heights, the union's underlying position was weak. The ASE had succeeded in organising only half of the fitters and turners in the industry, while the prior development of the division of labour had brought into the workshops an army of handymen capable of working the new machine tools who remained ineligible for union membership despite the efforts of the 1892 delegate meeting to widen the basis of recruitment. Whereas the compositors' successes in the 1910–11 strikes in Edinburgh and London depended on the support of the other printing unions, the ASE's history of demarcation disputes and organisational rivalries left it isolated in its confrontation with the employers, as the other major unions held aloof or deserted the joint committee at the onset of the lock-out. Ideologically and politically, the ASE was likewise on weak ground as the employers were able to represent them to public opinion as enemies of progress and property in the context of a wider legal and industrial offensive against the unions.[85]

The employers, by contrast, had achieved an unprecedented level of collective organisation and had been preparing for a national lock-out since the previous year. The Federation had been energetically seeking to win over the inland firms associated with the ITEA, and in the context of the intensification of conflicts over mechanisation and craft regulation, the eight-hours demand proved an ideal catalyst for the unification of the diverse sectors of the industry. A substantial reduction in working hours would at a stroke reduce productivity, raise costs, and undermine the competitive position of firms in all sections of the industry, and so rallied most major engineering employers behind the EEF's banner.[86]

At its peak, the lock-out involved some 45,000 men and some 25 per cent of ASE members, whose withdrawal was, however, insufficient to paralyse production completely. Some major firms ceased work completely, believing themselves dependent on an adequate quota of skilled workers; most continued operations at a reduced level, using black-legs supplied by professional strike-breaking organisations, and more importantly by promoting their labourers to skilled work. Thus while the ASE was able to raise a considerable strike fund, the financial strain of supporting the locked out men at a cost of £30,000 per week eventually became intolerable, and by November the union Executive was investigating peace terms.[87]

Whereas during the previous national lock-out in 1852 engineering employers had sought the outright destruction of the unions, by 1897 their strategy had shifted towards coercing the ASE Executive into forcing its members to abandon their local efforts at craft resistance. Thus while a public outcry and the resistance of ASE members persuaded the employers to offer guarantees of a role for collective bargaining in wage determination, the Terms of Settlement which concluded the disputes none the less embodied the key principles of managerial prerogatives. Employers were henceforth free to hire non-unionists; to institute piecework systems at prices agreed with individual workers; to demand up to 40 hours' overtime per man per month; to pay non-unionists at individual rates; to employ as many apprentices as they chose; and to place any suitable worker on any machine at a mutually agreed rate. In addition, the Terms of Settlement established a novel disputes procedure which prohibited strikes before a national conference had occurred between the union Executive and the EEF. In this way, the employers hoped to contain rank-and-file resistance to the reorganisation of the division of labour by deploying the constant threat of a national lock-out to force the ASE Executive to discipline its members.[88]

In the immediate aftermath of the 1897–8 lock-out, engineering employers believed that they had removed the major barrier to that transformation of the division of labour in their workshops which alone could repel the upsurge of foreign competition.[89] Yet by 1914 it had become clear that despite important gains, engineering employers had failed, at least in the older sectors of the industry, either fully to displace skilled workers from their central position in the division of labour or to break the back of craft resistance as a significant constraint on their freedom of action in the workplace. To be sure, engineering employers continued to introduce automatic and semi-automatic machine tools at as fast a rate as they had in the 1890s, and were able to undermine the long-term future of craft regulation through the increasing employment of semi-skilled and female labour; the multiplication of apprentices and the subversion of the technical content of their training; and the rapid extension of payment by results.[90] But ASE members were increasingly able to capture control of new machinery and 60 per cent of the workforce in Federated firms was still classified as skilled in 1914.[91] Productivity growth, as measured by throughput of metal per man hour, which had fluctuated around a rising trend in the 1880s and 1890s, levelled off after 1900, and foreign producers continued to expand their shares of world markets: by 1913 the UK exported £34.8 million of mechanical engineering products to £37.2 for Germany and £26.9 for the US.[92] Perhaps the most striking evidence of the limited transformation of the division of labour in this period can be found not in statistical series, but rather in the successes scored by a resurgent craft militancy from 1909 onwards, and in the continued dependence on skilled labour which emerged as a central aspect of the munitions crisis of the first world war.

Why did British engineering employers fail to take full advantage of the hegemonic position they had established in 1898 to effect a thoroughgoing transformation of the division of labour in the industry? The answer lies in a combination of the economic forces which shaped individual firms' investment strategies and the broader industrial relations strategy of the EEF; the intense commitment of ASE members and local officials to craft regulation and local autonomy; and the democratic character of the union which ensured that these views prevailed.[93]

Engineering manufacturers' approach to the division of labour in their works was largely determined by their investment strategies, which in turn depended in great measure on the nature of their product and the movement of demand. A full-scale transformation of the division of labour in line with the best practice engineering technology of the day involved extensive capital investment in new machinery, and often required a major reorganisation of workshop layout or even a purpose-built plant if it were to be fully effective. Thus such a transformation could be profitable only where there was a rapidly growing demand for a standardised, mass-produced product, which in Edwardian Britain was concentrated primarily in newer, lighter sectors such as cycles, motor cars, electrical goods and certain types of armaments.

While in these newer sectors British firms would be unable to enter world markets unless they adopted the most advanced methods, in those sectors where foreign competition remained ineffective before the war — principally textile engineering, but elsewhere as well — there was little incentive for manufacturers to undertake a wholesale transformation of production. In the short and medium term, British manufacturers could even concede a steadily growing share of European markets to their competitors if they themselves could maintain reasonable profit levels by turning to markets in the Empire and the rest of the underdeveloped world. Such tendencies were accentuated by the sharp shift of the terms of trade in favour of primary producers after 1900, which brought in its train a veritable Indian summer for the older sectors, as demand for their products rose to unprecedented levels.[94] At the same time, while the newer sectors grew rapidly up to 1914, prior investments in earlier types of infrastructure and the weakness of working-class purchasing power limited their expansion and profitability.[95]

In those large firms in the new sectors where rapidly rising demand facilitated major investment programmes, British manufacturers undertook rationalisations of factory layout and of the division of labour which pressed towards the limits of the new technology. In such cases, employers sought to appropriate to themselves and their supervisory staffs a greater share of the planning and direction of production, and to enforce tighter workshop discipline. A separate toolroom was often established where skilled workers designed the jigs and fixtures necessary for repetition production, and in many cases a new type of supervisor, the 'feed and speed' man was employed to set the optimum angles and speeds at which machines should be operated, usurping this traditional prerogative of the skilled worker. Where incentive bonus systems were in force, a rate fixer might also set output norms and piece prices based on primitive methods of work measurement. While these developments tended to displace the skilled worker from a direct role in production, it must still be noted that craftsmen were required in considerable numbers to make and set tools for the less skilled; to repair and maintain machinery, and even to perform production work where the nature of the task or the size of the run made mass-production methods impractical or uneconomic.[96]

In the bulk of the engineering industry, on the other hand, the small and unspecialised character of the firm, the structure of the market and the nature of existing plant discouraged major retooling, so that innovation consisted rather in the introduction of new machine tools and work practices within a workshop organisation that remained structurally unchanged. Indeed, a close examination of engineering employers' conduct in the aftermath of their

victory in 1898 suggests that their attempts to free themselves of craft regulation were more an extension of
traditional strategies for work intensification and cost
cutting than any breakthrough into a new rationalising and
Taylorist mode. The promotion of handymen to skilled
men's work, the extension of piecework and systematic
overtime, the subversion of apprenticeship into cheap boy
labour, together with questions of small wage advances and
reductions, thus dominated conflicts between skilled workers and their employers after 1898 as they had before. The
main novel element lay in employers' attempts to introduce
new systems of supervision and incentive payment (known
as premium bonus systems) designed to boost output and
reduce labour costs without major capital expenditure. Even
these grew out of methods already in force in some
engineering workshops before the lock-out, and tended in
practice to degenerate into rate-cutting exercises similar to
those used in conventional piecework.[97]

While individual employers pressed home their piecemeal
assault on craft regulation, responsibility for maintaining a
favourable overall framework for managerial prerogative
rested with the EEF, which relied principally on the new
disputes procedure inaugurated in 1898. Since no strike
could constitutionally take place until a deadlock had been
reached at national level, the Federation could use its
national strength to choke off local flare-ups of craft
militancy, as well as to stall district movements for wage
advances until after the peak of the trade cycle had passed.[98]

This strategy was reasonably effective so long as the ASE
Executive was prepared to work within the Terms of
Settlement. Conscious of the employers' dominant position
and of their determination to root out the remaining vestiges
of craft regulation, Barnes general secretary of the union,
sought to convert the demands of his members into a form
which could be made acceptable to the former by its
concentration on questions of remuneration. Hence Barnes
saw in the premium bonus a means of protecting engineering
workers against rate-cutting and of securing them a share of
the returns from increased output. As a socialist and new
unionist, Barnes urged the ASE to 'accept specialisation and

grade its members', and introduced a new section for the less skilled. As a determined centraliser, he welcomed the enhanced control over local militancy built into the new disputes procedure. At the same time, his stance involved no supine acceptance of the employers' interpretation of the Terms of Settlement, and ASE officials fought continuously for amendments which would alleviate rank-and-file grievances across a wide range of issues.[99]

Despite its dominance over union policy in the decade after 1898, Barnes' strategy ultimately foundered on the combination of the skilled engineers' tenacious defence of their craft status, the elaborate system of checks and balances in the ASE constitution, and the employers' determination to press forward a hard line on wages at the same time as a militant assault on craft regulation in the workplace. As their confidence returned in the years following the lock-out, ASE members in the districts resumed their pressure against the employment of handymen on machinery, enjoying considerable success where the firm involved was unwilling to refer the case to the disputes procedure. Similarly, the premium bonus, both because the practical shortcomings of the agreement negotiated by the union rendered it ineffective in restraining managerial abuses, and because the rank and file viewed the system as a radical threat to the principles of craft regulation.[100]

Rank-and-file opposition to Executive policy was generally supported by the union's own local officials. The district committees were directly responsible to the local membership, and the district delegates, though formally responsible to the Executive, were elected by the districts and tended in practice to represent local sentiment. Thus the districts were able to block the enrolment of the less skilled into the union, and local officials often placed themselves squarely behind strikes over machine manning and the premium bonus, even at the cost of suspension by the Executive.[101] More importantly, the principle of local autonomy was deeply entrenched in the ASE constitution, which defined the Executive as an administrative rather than policy-making body; control over policy was vested in the delegate meeting which controlled the formulation of union rules and in a Final Appeals Court responsible for their interpretation. The elective character of these latter bodies ensured their responsiveness to rank-and-file opinion, and a series of rule changes and Final Appeals Court judgments between 1896 and 1909 steadily limited the Executive's control over local strike movements.[102] The crucial turning points came with strikes against wage reductions on Clydeside in 1903 and on the north-east coast in 1908. The Executive's inability to prevent this latter strike provoked Barnes' resignation, and with it the collapse of his conciliatory strategy. Given the employers' determination to press forward a hard line both on wages and workshop organisation, there was little possibility of replacing the defence of craft regulation with economistic collective bargaining.[103]

With the erosion of Executive authority after 1908, and the rapid tightening of labour markets after 1911, craft militancy among ASE members enjoyed a dramatic resurgence. Its effects could be seen first of all in an intensified militancy in machine-manning disputes, which both mushroomed in number and proved vastly more successful for skilled workers than at any time since 1898. In several cases, large firms which had played a leading role in the offensive against craft regulation were forced to accept limitations on their rights to promote handymen to machines. This resurgence of craft militancy extended to the reorganisation of the division of labour more broadly, singling out the premium bonus for special opprobrium. While local militants were rarely able to dislodge the system once it had been introduced, they did succeed in blocking its extension to other major firms and in modifying the terms of its operation in a number of cases.[104]

By 1912, this revived craft militancy had captured the ASE Executive itself. After rule changes at the 1912 delegate meeting and a lengthy legal battle, the Old Executive was replaced by a new one committed to the vigorous defence of craft regulation. The first act of the new Executive was to ballot the membership on the abolition of the Terms of Settlement and the Carlisle Agreement on the premium bonus, proposals which obtained huge majorities from the membership. At the end of 1913, the Executive

issued three months' notice of its unilateral termination of
these agreements, and demanded the opening of negotiations for the 48-hour week. Constrained like its predecessor
to establish some *modus vivendi* with the powerful EEF, the
Executive concluded a new agreement with the employers
covering only the disputes procedure, and this was presented
to the membership as purely an interim measure.[105]

The resurgence of craft militancy from 1908 onwards
posed a major challenge to the tactics developed by the EEF
in the years following the lock-out. While long experience
under the disputes procedure and the difficulty of mounting
a national lockout at the apogee of a boom disposed the
EEF to keep open negotiations with the ASE Executive, the
Federation was at the same time strengthening its hand for a
possible confrontation by devising a compulsory strike
insurance scheme which came into effect in 1914. Engineering employers were not prepared to see all that they had won
in 1898 slip away, and it seems likely that a renewed
confrontation would have followed had the onset of war not
intervened.[106]

Despite the provisions of the Terms of Settlement, skilled
engineers' struggles to win control of new machinery had
proved remarkably successful during the decade and a half
before the first world war, a trend which became particularly
pronounced with the changes in the ASE Executive and the
tightening labour market after 1911. This short-term effectiveness of craft resistance cannot, however, obscure the
underlying erosion of craft regulation in the industry, a point
which emerges most clearly by comparison to the typographical unions' achievements during the same period.
Whereas the printing unions had secured virtually complete
control over composing machines, in engineering semi-
skilled labour was still expanding more rapidly than total
employment, though it is difficult to say by how much, and a
significant number of handymen were operating machines
similar to those worked by a growing number of ASE
members. The most rapidly growing sectors of the industry,
moreover, were those which disproportionately employed
the less skilled, such as cycles, motor cars, and electrical
engineering; in those regions such as the West Midlands

where the new trades predominated, the semi-skilled had begun to organise their own unions and erode the skilled men's wage differential during the 1911–14 labour unrest.[107]

Similarly, whereas the typographical unions had been able to bring under control the disadvantageous systems of payment and supervision which had played so large a part in depressing their members' earnings and working conditions before mechanisation, the continued progress of piecework and incentive payment marked the erosion of craft regulation in engineering. By 1914 some 46 per cent of fitters and 37 per cent of turners were paid by the piece, while nearly 10 per cent of ASE members were working under the hated premium bonus system.[108] While the revival of craft militancy helped curb the extension of the premium bonus and district committees and shop stewards gradually acquired a greater role in fixing piece prices, once installed the system was rarely dislodged even at the height of rank-and-file bargaining power on the eve of the war.

Perhaps the most important issue for the long-term future of craft regulation was the establishment of effective controls over apprenticeship, and it was here that the contrast between printing and engineering was sharpest. While the erosion of apprenticeship regulation was a central aspect of the deterioration of the hand compositor's position after the mid-century, its revival from the mid-1880s was both a contributory cause and a consequence of the successful struggle for control over composing machines, and the agreements signed in the 1890s proved the cutting edge of employers' growing formal acceptance of the unions' apprenticeship ratios. In engineering, on the other hand, apprenticeship was clearly in decay, as many employers recognised no restrictions and apprentices approached — or even exceeded — journeymen in number in some shops; in most cases apprentices were seen as a source of cheap labour and were often employed on repetitive processes to the detriment of their technical training. While the engineering unions sought in conferences with the employers immediately before the war to press their case for a revised apprenticeship system which would extend to machinists as well as to the older trades, the latter refused any concessions despite the upsurge of shop floor militancy.[109]

This underlying weakness of craft regulation in engineering ensured that the long-expected confrontation between skilled workers and employers would result in a resounding victory for the latter. The 1922 lock-out, unfolding against the background of large-scale unemployment, indeed resulted in a restoration of managerial prerogative in the workplace more complete than in 1898.[110] In the newer sectors, where the unions had been driven off the shop floor, the rapid growth of mass markets permitted the adoption of mass-production methods and a displacement of skilled workers along lines hinted at before the war.[111] In the older sectors, by contrast, the same economic forces responsible for the magnitude of the unions' defeat in 1922 meant that it would be mass unemployment rather than a further transformation of the division of labour which would strike most sharply at the position of the skilled worker.[112]

VI

Just as the absence of major technical innovation in printing and engineering between 1850 and 1890 did not render secure the position of skilled workers, so too the waves of mechanisation at the end of the century did not in themselves spell the demise of craft regulation. In printing, the typographical unions were able to take advantage of their early capture of composing machines to re-establish their framework of regulation on a firmer basis than in the hand era. Even in engineering, where craft regulation declined in the long term, skilled workers were able to defend their position in the division of labour with remarkable effectiveness up to 1914. When the distribution of skills shifted dramatically between the wars, the cause lay as much with the overall shifts in the importance of new and old sectors as with the reorganisation of work in the older trades themselves.

Why were compositors so much more successful than engineers in defending, and even enhancing, their position in the division of labour in the face of pressures towards technical and organisational change?[113] The conventional explanations which can be derived from the existing historiography of the two industries emphasise the roles of union policy, technology and market forces. But on closer inspection, none of these factors seems capable of accounting for the variations between the experiences of the engineers and the compositors. Thus historians of printing have explained the compositors' capture of mechanical typesetting primarily by reference to the judicious strategy adopted by their leaders towards the machines. As we have seen, however, while the typographical unions indeed displayed a shrewd tactical sense, the demands of the ASE were if anything more moderate and the policy of their Executive more cautious. Similarly, historians have stressed the skills technically required for the optimal operation of composing machines as an explanation for the employment of former hand compositors. While real skills were required of a composing machine operator, it is by no means clear that these were substantially greater than those of an ordinary typist, and they were certainly less than those of the badly paid case jobbing hand. Nor can it be said that the introduction of automatic machine tools eliminated at a single stroke the need for skilled craftsmen in engineering; as we have seen, a substantial number of skilled workers were required in even the most advanced firms to perform more highly skilled work connected with mass production. Most tellingly, however, the employment of women on monotypes strongly supports the view that from a technical point of view composing machines would have been compatible with a radically different division of labour, modelled perhaps on the Edinburgh pattern, in which a small number of fully-trained compositors were retained for highly skilled tasks, while the bulk of the typesetting was performed by women after a relatively short training period. All the evidence suggests that it was only under intense union pressure that printing employers were prepared to replace women with men as monotype operators.

As the case of the monotype suggests, the divergent abilities of engineers and compositors to control the impact of technical change on their position in the division of labour can better be explained by variations in the patterns and outcome of industrial conflict. At issue here were both the outcomes of full-scale confrontations between unions and employers, and the less dramatic evolution of the balance of industrial power through local skirmishes and collective bargaining. A crucial determinant of the balance of forces between skilled workers and employers in each case lay in the relative cohesion and capacity for collective action of each side. Thus employers in engineering were both more aggressive and significantly more unified than their counterparts in printing, while the reverse was true of the workers in the two industries. In engineering, the employers seized on the arrival of a renewed wave of mechanisation in an attempt to free themselves from the restrictions of craft regulation, and created a powerful organisation able to formulate policy on a centralised basis. In printing, on the other hand, newspaper proprietors conceded initial control over the linotypes to skilled compositors without a struggle, and particularly in London left the task of resisting the demands of the printing unions to the large book firms. On the workers' side, the ASE in 1897–8 failed to receive the support of the other main craft unions in the industry, and had earned the active hostility of the unskilled, whether organised or not. By contrast, the movement led by compositors for the 50-hour week in London and the elimination of underpaid female labour in Edinburgh were strongly supported not only by the other printing crafts, but also by the new unions of the less skilled, whose cooperation was by all accounts crucial to their success.

These variations in solidarity and bargaining power of skilled workers in the two industries can be ascribed in the first instance to structural factors. Thus the variations in employers' bargaining postures and capacities for collective action stemmed in large part from differences in their market position. The sharpest contrast is that between engineering and newspaper printing. The newspaper proprietors were on the one hand specially vulnerable to strike action because of the perishable character of their product and sharply divided among themselves by competition for a highly volatile readership, and on the other able to concede union demands because of the underlying buoyancy of their economic position. While engineering firms were less fragmented by domestic competition, unlike the newspaper and jobbing printers, they faced growing competition from abroad, which together with the slow growth of demand for their products deterred them from seeking compromises with skilled workers over the introduction of new techniques. This general contrast is underlined by the distinctive position of the large book firms within printing itself. Able to stockpile their relatively standardised and durable product, but vulnerable to competition from provincial plants enjoying non-union wage rates and work rules, it was the book firms which formed the backbone of employer militancy in British printing and which made the greatest efforts to dispense with skilled operators on composing machines.

Similarly, the variations in the solidarity of engineering and printing workers were closely related to the sharply different social relations prevailing between sections of the labour force in the two industries, which were in turn rooted in structural differences in the divisions of labour prevailing before mechanisation. Paradoxically, the more amicable relations between craftsmen and the less skilled in printing resulted from the greater success of the compositors' exclusiveness. The static technology in the composing room meant that there were no non-craftsmen with a foothold in the organisation of production who could be upgraded by the introduction of the linotype, whereas in engineering a previous wave of mechanisation had called into being a host of handymen who could expect promotion to the new machine tools unless restrained by the ASE. At the same time, while there was little overlap between compositors' work and that claimed by other printing trades, the heterogeneous structure of the engineering industry and the greater complexity of the division of labour rendered skilled engineers' demarcation lines ambiguous and difficult to police, bringing them into recurrent conflict with other trades, especially in the shipyards.

While market structure and the division of labour account for much of the divergence in the behaviour of skilled workers and employers in the two industries, they do not provide a complete explanation. While structural forces facilitated certain alliances and discouraged others, the actual coalitions forged by skilled workers and employers depended in large measure on those combinations of conjunctural circumstances, past experience of industrial conflict, and strategic choices which made them appear necessary and desirable to the participants. Thus the initial stances adopted by printing and engineering employers towards mechanisation, which significantly influenced the success of craft regulation, grew out of partially mistaken assumptions about its implications for the future of the division of labour which were shaped by previous experiments with technical change. Structural forces can moreover be discerned which acted to obstruct as well as encourage the formation of a cohesive employers' organisation in engineering, such as the divergence of interests between firms in diverse product markets on which the ITEA ran aground, and the formation of the EEF depended on the convergence of intensified conflict over routine issues, foreign competition and new opportunities for mechanisation to overcome the sectional barriers among engineering employers. A similar convergence of unrelated struggles made possible the grand coalition of provincial printing employers which averted a national strike in 1911, and it is by no means inconceivable that even the London newspaper proprietors could have been forced into collective action by tighter union restrictions on composing machines.

On the workers' side, the construction of a durable and effective Printing Trades Federation, though facilitated by the amicable relations between compositors and other trades, was the product of a lengthy and eminently political process of bargaining among the various unions in the industry, which was then solidified by the experience of joint struggles after 1900. Conversely, the ASE's aggressive and autocratic political style did much to worsen its relations with other skilled trades, while the failure of the various attempts to build an alliance between craftsmen and the less skilled from 1892 to 1914 was the result not only of rank-and-file opposition, but also of the constitutional arrangements that allowed this opposition to determine union policy. Thus the alliances and cleavages within the ranks of capital and labour alike which exerted so strong an influence on the outcome of industrial conflict and the future of craft regulation can best be understood as the product of a dialectic between such structural factors as market position and the shape of the existing division of labour, and the creative response of each group of actors to the dangers and opportunities presented by any historical situation.

NOTES

I am pleased to acknowledge the generous access to archival material provided by the Engineering Employers' Federation.
All works printed in London unless otherwise indicated.

1. E.J. Hobsbawm, 'The Labour Aristocracy in 19th Century Britain', in *Labouring Men* (1964); R.Q. Gray, *The Labour Aristocracy in Victorian Edinburgh* (Oxford, 1976).
2. H. Pelling, 'The Concept of the Labour Aristocracy' in *Popular Politics and Society in Late Victorian Britain* (1968); J. Foster *Class Struggle and the Industrial Revolution* (1974); G. Stedman Jones, 'Class Struggle and the Industrial Revolution', *New Left Review* 90 (1975). A fuller bibliography of the labour aristocracy debate will be found in the introduction to this volume.
3. For the principal components of the new synthesis emerging in opposition to H. Braverman's *Labour and Monopoly Capital* (New York, 1974), see S. Wood (ed.) *The Degradation of Work? Skill, Deskilling and the Labour Process* (1982); C. Sabel, *Work and Politics* (Cambridge, 1982); and J. Zeitlin, 'Social Theory and the History of Work', *Social History* 8 (3) 1983.
4. Hobsbawm, *op. cit.*, pp.283–4, 288, 301; Stedman Jones, *op. cit.*, p.63; *cf.* also K. Burgess, *The Origins of British Industrial Relations* (1975), ch.1.
5. Pelling, *op. cit.*, p.45; Hobsbawm, *op. cit.*, p.280; Gray, *op. cit.*, pp.57–60, 89.
6. A.J. Lee, *The Origins of the Popular Press, 1855–1914* (1976); J. Child, *Industrial Relations in the British Printing Industry* (1967), p.107.
7. A.E. Musson, 'The Newspaper Industry in the Industrial Revolution', *Economic History Review*, 2nd ser., X (1958); B.W.E. Alford, *The London Letterpress Printing Industry, 1850–1914*, (PhD thesis, London, 1962), ch.2.
8. J. Southward, *Practical Printing* (1882); and 'Type-Composing Machines of the Past, Present and Future' (unpublished paper, 1890,

Southward Collection, St Bride's Library); J.H. Zeitlin, *Craft Regulation and the Division of Labour: Engineers and Compositors in Britain, 1890–1914*, PhD thesis, Warwick, 1981 (hereafter referred to as Zeitlin, *Thesis*), pp.35–38.

9. Child, *op. cit.*, chs 7–9) Musson, *The Typographical Association* (Oxford, 1954), chs 9–10; Zeitlin, *Thesis*, chs 2–3.
10. Ibid., pp.17–31.
11. Ibid., pp.134–7.
12. *Scottish Typographical Circular (STC)*, December 1890.
13. LSC submission to the Royal Commission on the Depression of Trade and Industry, *Second Report*, Parliamentary Papers 1886, XXII, Appendix II, p.79; F. Willis, *The Present Position And Future Prospects of the LSC* (1881); 'High Pressure', *STC*, March 1885.
14. *Typographical Circular (TC)*, August 1892, p.6.
15. Zeitlin, *Thesis*, pp.6–7, 143.
16. Musson, *op. cit.* (1954), p.103; *TC* January 1893, p.4; Wages Census, 1906, vol.VII, *Printing and Paper Trades*, Parliamentary Papers 1913, CVIII, pp.29ff.
17. LSC, 'Report of a Sub-Committee on Systematic Overtime', *Trade Reports* 1892; Zeitlin, *Thesis*, pp.147–8.
18. Ibid., pp.135–9, 146–8.
19. Ibid., pp.139–42.
20. E. Edwards, *The Disease and the Remedy: An Essay on the Distressed State of the Printing Trade, Proving it to be Mainly Attributed to Excessive Boy Labour* (1850); Zeitlin, *Thesis*, pp.149–51 and Table 7A.
21. Ibid., pp.151–2 and Tables 7B and 7C.
22. S. and B. Webb, *Industrial Democracy* (1902), pp.464–8; S.C. Gillespie, *The Scottish Typographical Association, 1853–1952* (Glasgow, 1953), pp.96–8; Child, *op. cit.*, pp.132–6; Zeitlin, *Thesis*, pp.152–4.
23. For membership figures, see E. Howe and H.E. Waite, *The London Society of Compositors* (1948), p.338; Musson, *op. cit.*, (1954), p.535; STA, *Annual Reports*, 1880, 1893. For estimates of union density, see Musson, *op. cit.* (1954), p.115; and Zeitlin, *Thesis*, pp.156–7.
24. Child, *op. cit.*, p.110; A. Linnett, 'Women Compositors', *Economic Journal*, January 1892; 'Statement of the Edinburgh Branch of the Female Question', *STC*, September 1904; Fair Wages Committee, *Minutes of Evidence*, Parliamentary Papers, 1908, XXXIV, testimony of T.E. Naylor (LSC), G. Templeton (STA), W. Fraser (Neill and Co., Edinburgh), and T. Richards (Thom and Co., Dublin); Zeitlin, *Thesis*, pp.157–60.
25. Ibid., pp.160–2.
26. A.L. Bowley and G. Wood, 'The Statistics of Wages in the 19th Century, pts V, (Printers) and X–XIV (Engineering and Shipbuilding Trades)', *Journal of the Royal Statistical Society*, LXII (1899) and LXVIII–LXIX (1905–6); E.H. Phelps Brown and M. Browne, *A*

Century of Pay (1968), p.358; LSC memorial 1890 in E. Howe (ed.), *The London Compositor* (1947), pp.317–9; Zeitlin, *Thesis*, pp.162–4.

27. *London Typographical Journal* (*LTJ*), May 1906, p.5; 'Is the Printer Deteriorating?', *STC*, August 1886; I.C. Cannon, *The Social Situation of the Skilled Worker*, (PhD thesis, London 1961), pp.86–8; Zeitlin, *Thesis*, pp.165–6.

28. K. Burgess, *op. cit.*, pp.5–24; idem, 'Technological Change and the 1852 Lock-out in the British Engineering Industry', and 'Trade Union Policy and the 1852 Lock-out in the British Engineering Industry', *International Review of Social History*, XIV (1969) and XVII (1972).

29. Foster, *op. cit.*, p.227; cf. also Stedman Jones, *op. cit.*

30. J.B. Jefferys, *The Story of the Engineers* (1946), esp. p.16; Zeitlin, *Thesis*, pp.41–8.

31. Burgess, *op. cit.*, (1975), esp. pp.25–8; Jefferys, *op. cit.*, p.291; on specialisation, Zeitlin, *Thesis*, pp.48–50.

32. M. and J.B. Jefferys, 'The Wages, Hours and Trade Customs of the Skilled Engineer in the 1861', *Economic History Review*, 1st ser., XVII (1947); Burgess, *op. cit.*, (1975), pp.24–41; Zeitlin, *Thesis*, pp.96–102.

33. Burgess, *op. cit.* (1975), p.41; J. Burnett, *The Nine-Hours Movement: A History of the Engineers' Strikes in Newcastle and Gateshead* (Newcastle, 1872); E. Allen *et al.*, *The North-East Engineers' Strikes of 1871* (Newcastle, 1971).

34. J. Price (General Manager, Palmers' Engineering and Shipbuilding Works, Jarrow) to Royal Commission on the Depression, *op. cit.*, Appendix IV, and *Third Report*, Parliamentary Papers 1886, XXIII, q. 10,963–7; W.A. Lewis, *Growth and Fluctuations, 1870–1913* (1978), pp.34–57; S.B. Saul, 'Engineering', in D.H. Aldcroft (ed.), *The Development of British Industry and Foreign Competition, 1875–1914* (Glasgow, 1968A); R. Floud, 'The Adolescence of American Engineering Competition, 1860–1900', *Economic History Review*, XVII (1974); I.W. McLean, 'Anglo–American Engineering Competition, 1870–1914: Some Third Market Evidence', *Economic History Review*, XXIX (1976).

35. Submission by ASE to Royal Commission on Labour, Group A, *Minutes of Evidence*, Parliamentary Papers, 1893–4, XXXII, Appendix XLVI; Jefferys, *op. cit.*, pp.100–1; Burgess, *op. cit.*, pp.39–40, 45–6; Zeitlin, *Thesis*, pp.105–8 and Table 4.

36. Price to Royal Commission on the Depression, Third Report, q. 10,963.

37. Saul, 'The American Impact on British Industry, 1895–1914'. *Business History*, III (1960); 'The Machine-Tool Industry in Britain to 1914', *Business History*, X (1968B); idem, *op. cit.*, (1968A); Floud, *op. cit.*, Price to Royal Commission on the Depression, *Second Report*, Table 1, and *Third Report*, q. 10,971; E. Harland (Harland and Wolff, Belfast) to Webbs, Webb Trade Union

Collection, EA XV, f. 1; ibid. XVI, f. 6, pp.54–61, 63–8; Zeitlin, *Thesis*, pp.110–13.

38. Royal Commission on Labour, op. cit., Appendix XLVI and testimony of ASE witnesses in *ibid.*; Webb Trade Union Collection, EA XVI, f. 6; Zeitlin, *Thesis*, pp.113–7.

39. Col. H. Dyer (managing director) to B. Potter, 15 October 1891. Webb Trade Union Collection, EA XXI, f. 18; ASE, *Quarterly Report*, June 1894. For the connection between the use of handymen and repetition production at Armstrongs, see the testimony of Sir B.C. Browne to the Royal Commission on the Poor Laws, *Minutes of Evidence*, Parliamentary Papers 1910, XLVIII, q. 86,227.

40. Captain A. Noble (managing director) to Committee on the Organisation and Administration of the Army Manufacturing Departments, *Minutes of Evidence*, Parliamentary Papers 1887, XIV, esp. qq. 8998–9016; W.G. Gordon, *Foundry Forge and Factory* (1890), p.34; D.F. Schloss, *Methods of Industrial Remuneration* (1898 edn), p.15; J. Ratcliffe (ASE) to Select Committee on Government Contracts, *Minutes of Evidence*, Parliamentary Papers 1897, X, qq. 2468, 2506, 2536; B. Taylor, 'The Machine Question and Eight Hours', *Cassier's Magazine*, November 1897; for a similar description of these supervisors from a trade union perspective, see the testimony of W. Glennie (ASE) to Royal Commission on Labour, *op. cit.*, q. 23,157. I am indebted to Keith McClelland for several of these references.

41. Jefferys, *op. cit.*, p.119; Royal Commission on Labour, *op. cit.*, Appendix XLVI; 1886 Wages Census, Parliamentary Papers 1893–4, LXXIII; Zeitlin, *Thesis*, pp.117–19, 128–9 and Table 5. This process was particularly evident in depressions: thus in 1878 several Glasgow firms were paying skilled engineers only 22–4 shillings per week as opposed to the district rate of 27s 6d, while the following year the ITEA announced that the ten hour day had been reimposed in ten of its seventeen districts without a struggle. *Capital and Labour*, 31 July 1878 and 21 May 1879, quoted in K. Burgess. 'The Amalgamated Society of Engineers before 1914: An Old or New Union?' in W.J. Mommsen and H.G. Husung (eds.), *Towards Mass Unionism: Trade Unions in Great Britain and Germany, 1880–1914* (forthcoming, 1985).

42. On union density, see Burgess, *op. cit.* (1975), p.82, and the estimates collected in Zeitlin, *Thesis*, pp.127–8.

43 Ibid., pp.122–7; A.J. Reid, *The Division of Labour in the British Shipbuilding Industry, 1880–1920*, (PhD Thesis, Cambridge, 1980), pp.91–5, 100–2, 150–4; P. Robertson, 'Demarcation Disputes in British Shipbuilding to 1914', *International Review of Social History*, XX (1975).

44. Zeitlin, *Thesis*, pp. 116–19, 180–7.

45. Musson, *op. cit.* (1954), pp.99–101, 221–4; and (1958); Southward, *op. cit.* (1890), and his *Progress in Printing and the Graphic Arts during the Victorian Era* (1897); J.S. Thompson, *A History of*

Composing Machines (Chicago, 1904); L.A. Legros and J.C. Grant, *Typographical Printing Surfaces* (1916); Zeitlin, *Thesis*, pp.209–15.

46. See the technical sources cited in the previous note.

47. TA survey, reprinted in Howe, *op. cit.* (1947), pp.498–9; Linotype Co., *Report of Directors to the 7th Annual Meeting of Shareholders* (1896), Webb Trade Union Collection, EB LXXIV, f. 61.

48. 'Report of the Meeting of Newspaper Proprietors and Printers with London Members of the LUA, 7 November 1895', Webb Trade Union Collection, EB LXXIV, f. 57, especially the remarks of Sir E. Lawson (*Daily Telegraph*), p.10; G.E. Hart (*Financial Times*), 'The Trouble in the Printing Trade', *Magazine of Commerce*, June 1903; Linotype Co. quoted in Webbs, *op. cit.*, p. 407; Zeitlin, *Thesis*, pp.218–21.

49. TA, *Report of a Conference of Composing Machine Operators, Apr. 1893* and *Report of a Special Delegate Meeting, Dec. 1893*; TA *Executive Minutes* 1893–5, *passim*; *British and Colonial Printer and Stationer* (BCPS) 19 May 1895; Musson, *op. cit.* (1954) pp.224–35; Zeitlin, *Thesis*, pp.222–30.

50. Linotype Co., 'To Unemployed Young Men', reprinted in *TC*, October 1894; 'Report of Royal Commission Meeting', *TC*, October 1894; TA Executive, 'Report on Composing Machines', quoted in 'MS. Report to a Special General Meeting, 1 January 1896', LSC, *Special Reports* (Modern Records Centre); TA, *Executive Minutes* 1895–8; Musson, *op. cit.* (1954), pp.230–7; Zeitlin, *Thesis*, pp.233–4, 239–40; the 1898 agreement is reproduced in *TC*, January 1899.

51. *TC* (1899), *passim*; *Report of RC Meeting, 1899*; Zeitlin, *Thesis*, pp.235–41.

52. On the reformation of the London MPA, see *BCPS*, 11 April 1890, reprinted in LSC, *Trade Reports* (1890); the 1894 agreement is reproduced in Howe, *op. cit.*, pp.497–501; the quotation is from 'MS. Report to a Special Delegate Meeting, 14 December 1895', LSC, *Special Reports*, p.3; Zeitlin, *Thesis*, pp.24–7, 243–4.

53. 'Report to a Special Delegate Meeting, 14 December 1895; 'Report of the Meeting of Newspaper Proprietors . . . 7 Nov. 1895'; circular in Webb Trade Union Collection, EB LXXIV, f. 40; revised scale reprinted in Howe, *op. cit.*, pp.504–6.

54. For the contrast between the internal government of the two unions, see Zeitlin, *Thesis*, pp.81–7.

55. Accounts of meetings in LSC, *Trade Reports* and *Special Reports* 1894–6, *passim*; Zeitlin, *Thesis*, pp.247–50.

56. 'Synopsis of Opinions and Suggestions during the Past Fortnight by American News Editors and Managers Passing through London *re* the Agreement between the London Morning and Evening Papers and the LSC which Expires 31 December 1895' (n.d.), Webb Trade Union Collection, EB LXXIV, f. 33; 'MS. Report of an Interview with Sir E. Lawson, 23 January 1896', LSC *Special Reports*; 'Report of the Lintotype Co. Shareholders' Meeting, 25 January 1896', Webb Trade Union Collection, EB LXXIV, f. 60, p.25; *BCPS*, 25

December 1895; G.E. Hart, *The Linotype: A Comparison of Cost* (1908); LSC, *Trade Reports* (1896); revised agreement reprinted in Howe, *op. cit.*, document CXXIV; Zeitlin, *Thesis*, pp.250–3.

57. STA, *Annual Reports* (1896–9); Zeitlin, *Thesis*, pp.253–9.
58. TA, *Executive Minutes* (1899–1903), *passim*; *TC*, (1899–1903), *passim*; LUA, *Monthly Circulars*, (1899–1903), *passim*; Musson, *op. cit.* (1954), pp.237–42; Zeitlin, *Thesis*, pp.283–8.
59. LSC, *Trade Reports* and *Special Reports* (1899–1903), *passim*; LUA, *Monthly Circulars*, (1899–1904), *passim*; Hart, *op. cit.* (1903); Zeitlin, *Thesis*, pp.267–70, 272–5.
60. BCPS, 11 April 1890; *Notes of the Proceedings of an Arbitration between the LSC and the London MPA before G.R. Askwith, 11 February 1901*; Alford, *op. cit.*, pp.50–63; J.S. Elias, *The Monotype from a Printer's Point of View* (1908); LSC, *Trade Reports*, (1902–5), *passim*; Child, *op. cit.*, pp.181–2.
61. Musson, *op. cit.* (1954), pp.241–6; Zeitlin, *Thesis*, pp.288–90.
62. Report of LUA Annual General Meeting 1904, in *TC*, June 1904; testimony of S.C. Straker (Straker Bros., London) to Fair Wages Committee, *op. cit.*, q. 4485; LSC, *Annual Report* (1904), p.32; C.J. Bundock, *The Story of the National Union of Printing, Bookbinding and Paper Workers* (Oxford, 1959), pp.143–4; TA, *Report of Delegate Meeting 1903*, p.29.
63. STA, *Annual Reports* (1904, 1905, 1909–12); *Scottish Typographical Journal (STJ)* (1910), *passim*; Gillespie, *op. cit.*, pp.204–7; Zeitlin, *Thesis*, pp.321–8.
64. Ibid., pp.293–8, 301–3.
65. LSC, *Trade Reports* (1907–9), *passim*; *Report of TA Delegate Meeting* (1908); Zeitlin, *Thesis*, pp.279–82, 290–2, 301–2.
66. Statement issued by joint conference of provincial employers associations, Master Printers' Association, *Members' Circular*, September 1910; Zeitlin, *Thesis*, pp.302–5. An additional factor dividing the LSC and the TA lay in the conflict arising from the former's decision unilaterally to extend its radius from 15 to 40 miles from the centre of London in 1907; a TUC-sponsored amalgamation proposal foundered on the LSC's insistence on local autonomy in trade policy, leaving a heritage of bitterness in the provinces. See Zeitlin, *Thesis*, pp.300–2.
67. LSC, *Annual Report* (1911); 'Review of the Year 1911', *Printers' Register* January 1912; Zeitlin, *Thesis*, pp.307–15, 317–20 and Table 8.
68. *TC*, June 1911, p.7; MPA, *Members' Circular*, April 1911; Musson, *op. cit.* (1954), pp.167–9, 192, 197, 216, 248–9, 296–7; TA, *Report of Delegate Meeting* (1913); STA, *Annual Reports* (1911–13); *STJ* (1912–13), *passim*; Child, *op. cit.* pp.228, 262–7; Zeitlin, *Thesis*, pp.315–8, 328–30, 479–80.
69. Board of Trade, *Report of an Enquiry into the Conditions of Apprenticeship and Industrial Training* (1915), pp.242–3, 252; Ministry of Labour, *Report on Apprenticeship and Training, 1925–6*,

vol.I, pp.9–16.
70. W.H. Beveridge, *Unemployment: A Problem of Industry* (1908), pp.140–1; Alford, *op. cit.*, p.222; 'Report on Casual Labour and Piece-Stab', LSC *Trade Reports* (1909); Zeitlin, *Thesis*, pp.333–5, 485–8 and Table 14.
71. For the capacities of the new technology, see N. Rosenberg, 'Technological Convergence in the American Machine Tool Industry, 1840–1910', *Journal of Economic History*, XXIII (1963); Saul, *op. cit.* (1960), esp. p.22; and R. Floud, *The British Machine Tool Industry, 1850–1914* (Cambridge, 1976), ch.2. For the limited impact on British industry before the 1890s, see Saul, *op. cit.* (1960, 1968A and B) and his 'The Market and the Development of Mechanical Engineering, 1870–1914', *Economic History Review*, XXVII (1967).
72. Saul, *op. cit.* (1960, 1968B); Floud, *op. cit.* (1974); A.E. Harrison, 'The Competitiveness of the British Cycle Industry, 1890–1914', *Economic History Review*, XXII (1969); J. Blackman and E.M. Sigsworth, 'The Home Boom of the 1890s', *Yorkshire Bulletin of Economic Research*, XVII (1965).
73. Saul, *op. cit.* (1968B), p.29; Blackman and Sigsworth, *op. cit.*, pp.85–6.
74. See ASE *Annual Reports* (1887, 1892 and 1894) for evidence of national coordination of this campaign.
75. On London (Maxim–Nordenfeldt dispute), see ASE *Monthly Report* December 1889; *Abstract of Proceedings of ASE Local Executive Council* (1888–90), pp.18–24; B.M. Weekes, *The Amalgamated Society of Engineers, 1880–1914*, PhD thesis, Warwick, 1970, ch.1. On Lancashire, P. de Rousiers, *Le trade-unionisme en Angleterre* (Paris, 1897), p.281. On the Bristol Channel, Webb Trade Union Collection, EX XVI, f. 6, pp.42–53; ASE, *Quarterly Report*, September 1894. On Barrow, *ibid.*, June 1894; de Rousiers, *loc. cit.*
76. ASE, *Monthly Reports* February 1889, April 1890; testimony of Glennie (ASE), Price (Palmers) and G. Cherry (Operative Society of Plumbers) to Royal Commission on Labour, *op. cit.*
77. Weekes, *op. cit.*, chs 2–3; Zeitlin, *Thesis*, pp.181–7.
78. E. Wigham, *The Power to Manage* (1973), ch.1; testimony of W. Mosses (Patternmakers), Whittaker (ASE) and Noble (Armstrongs) to Royal Commission on Labour, *op. cit.*; ASE, *Monthly Report*, February 1889; details of insurance scheme in letter from B.C. Browne to A. Smith, 15 May 1912, in EEF Archives I(4)1.
79. Board of Trade Labour Department, 'Report on Strikes and Lock-outs in 1895', Parliamentary Papers 1896, LXXX, pt. I, pp.32–6; Wigham, *op. cit.*, pp.22–4; Weekes, *op. cit.*, p.83; Jefferys, *op. cit.*, pp.140–1.
80. EEF, 'Conditions of Federation' (1896), reprinted in Wigham, *op. cit.*, Appendix B.
81. ASE, *Annual Report* (1896) and *Monthly Journal and Report*, January 1892, p.40; Letter from EEF to ASE, quoted in Weekes, *op. cit.*, p.84; EEF, *Executive Minutes*, 26 November 1896; and

Wigham, *op. cit.*, p.33.

82. EEF, *Executive Minutes*, 24 August 1896; Weekes, *op. cit.*, p.83; Wigham, *op. cit.*, pp.32–3.
83. EEF, *Executive Minutes*, 13 August 1896 and 12 March 1897; *Verbatim Report of a Conference between the ASE and the EEF on the Machine Question*, April 1897.
84. ASE, *Monthly Journal and Report*, June–July 1897; ASE, *Notes on the Engineering Trades Lockout* (1898); Wigham, *op. cit.*, pp.38–43; Weekes, *op. cit.*, pp.190–2.
85. Jefferys, *op. cit.*, p.292; Burgess, *op. cit.* (1975), p.47; ASE, *op. cit.*, pp.4, 20–7; R.O. Clarke, 'The Dispute in the British Engineering Industry, 1897–8: An Evaluation', *Economica*, May 1957; Zeitlin, *Thesis*, pp.197–200.
86. EEF, *Executive Minutes* 24 April and 26 November 1896, 21 March 1897; *List of the Federated Engineering and Shipbuilding Employers who Resisted the Demand for a 48-Hours Working Week, 1897–8* (Glasgow, 1898); EEF, *Executive Reports* (1897–8), *passim*.
87. B.C. Browne to Royal Commission on Trades Disputes and Combinations, *Minutes of Evidence*, Parliamentary Papers 1906, LVI, q. 2574; EEF, *Executive Report*, 35, 12 August 1897; Wigham, *op. cit.*, Appendix C. For fuller accounts of the lock-out, see *inter alia* Board of Trade Labour Department, 'Report on Strikes . . . in 1897', Parliamentary Papers 1898, LXXXVII, pp.1ii–lx; Clarke, *op. cit.*; Weekes, *op. cit.*, ch.4; Wigham, *op. cit.*, ch.2; Jefferys, *op. cit.*, pp.144–8; H. Clegg, A. Fox and A.F. Thompson, *A History of British Trade Unions, 1889–1910*, vol.1 (Oxford, 1964), pp.161–8.
88. For employer attitudes and the debate over the threat to collective bargaining, see ASE, *op. cit.*, pp.74–5, 116–26, 133–7, 148–52; Clegg, Fox and Thompson, *op. cit.*, pp.164–7; Zeitlin, *Thesis*, pp.202–6. The Terms of Settlement are reprinted in Wigham, *op. cit.*, Appendix D.
89. L. Cassier, 'The British Engineers' Strike of 1897–8: Its Lessons and Results', *Cassier's Magazine*, April 1900; J. Slater Lewis, 'Works Management for Maximum Production', *Engineering Magazine*, May 1900; and B.C. Browne, 'Uses and Abuses of Organisation among Employers and Employed', *ibid.*, January 1901.
90. See Zeitlin, *Thesis*, pp.433–7 for details.
91. M.L. Yates, *Wages and Labour Conditions in British Engineering* (1937), p.31.
92. Phelps, Brown and Browne, *op. cit.*, pp.174–95, esp. pp.177, 180–1; Saul, *op. cit.* (1968A).
93. For a fuller treatment of employer strategies, see my 'The Labour Strategies of British Engineering Employers, 1890–1922', in H. Gospel and C. Littler (eds), *Managerial Strategies and Industrial Relations: An Historical and Comparative Study* (1983).
94. Saul, *op. cit.* (1967 and 1968A); Lewis, *op. cit.*, esp. chs 4–5. For a similar overall argument, see E.J. Hobsbawm, *Industry and Empire* (Harmondsworth, 1969).

95. For the growth of the new sectors, see Jefferys, *op. cit.*, p.120; Saul, 'The Motor Industry in Britain to 1914', *Business History*, V (1962). For obstacles to their development, see I. Byatt, 'Electrical Products', in Aldcroft, *op. cit.*, and *The Electrical Industry in Britain, 1875–1914* (Oxford, 1979), ch.8; R.E. Caterall, 'Electrical Engineering', in N. Buxton and D.H. Aldcroft (eds), *British Industry between the Wars* (1979); Saul, *op. cit.* (1960 and 1967); R.J. Irving, 'New Industries for Old? Some Investment Decisions of Sir W.G. Armstrong, Whitworth and Co., 1900–14', *Business History*, XVII (1975).

96. For broad descriptions of the implications of the new techniques, see D.S. Landes, *The Unbound Prometheus* (Cambridge, 1972), pp.292–323; Jefferys, *op. cit.*, pp.124–5; Saul, *op. cit.* (1960), pp.28–9; Weekes, *op. cit.*, ch.5; J.W.F. Rowe, *Wages in Practice and Theory* (1928), Appendix III; and C. More, *Skill and the English Working Class, 1870–1914* (1980), pp.184–92.

97. This analysis is based on detailed case files in the EEF archives and the reports of ASE organising district delegates in the union's *Monthly Reports*. See Zeitlin, *Thesis*, pp.357–71.

98. See Zeitlin, 'Labour Strategies', pp.27–30.

99. G.N. Barnes, 'The Uses and Abuses of Organisation among Employers and Employed', *Engineering Magazine*, January 1901; 'Request by the ASE, SEMS and UMWA for Amendment of the Terms of Settlement', EEF, *Executive Report* 229, 4 April 1900; *Verbatim Report of Conferences between the EEF and the ASE, SEMS and UMWA, Dec.–May 1900*; transcripts of further conferences 1906–7, EEF Archives, A(2)5–9; Weekes, *op. cit.* pp.246–50; Zeitlin, *Thesis*, pp.372–406.

100. ASE, ODD reports in the union's *Monthly Journal and Reports* (1898–1904), *passim*; Zeitlin, *Thesis*, pp.376–7, 387–400.

101. As in Manchester in 1906 and Erith in 1909. On Manchester, ASE, *Monthly Reports* September, November 1906; EEF, *Decisions of Central Conference, 1898–1925* (1925), case 1247, p.259; EEF, Archives M(5)1. On Erith, see TUC Joint Committee, *Report on the Premium Bonus System* (Manchester, 1910), pp.50–3; Weekes, *op. cit.*, pp.185–200.

102. On the constitutional structure of the ASE, see ibid., ch.1; and Zeitlin, *Thesis*, pp.78–81.

103. R. Croucher, *The ASE and Local Autonomy, 1898–1914*, MA thesis, Warwick, 1971; Weekes, *op. cit.*, ch.6; Zeitlin, *Thesis*, pp.401–5.

104. Zeitlin, *Thesis*, Tables 10–11; EEF, Archives M(6)8; EEF, *Executive Report for 1913*; EEF, *op. cit.* (1925), case 306, p.68, case 681, p.131, and cases 1436 and 1439, pp.298–300; TUC Joint Committee, *op. cit.*; S. Pollard, *A History of Labour in Sheffield* (Liverpool, 1959), p.232; ASE, *Monthly Journal and Report*, January 1913.

105. Jefferys, *op. cit.*, pp.169–71; Weekes, *op. cit.*, ch.8; 'Verbatim Transcript of Special Central Conference between the ASE and the

EEF, 13 February 1914', EEF Archives, A(4)6; EEF, *Executive Minutes* 7 February 1913, 16–17 April, 1914. For the text of the new agreement, see A.I. Marsh, *Industrial Relations in Engineering* (Oxford, 1965), Appendix C3.

106. EEF, *General Letters*, 171, 175, 176, 181, 30 April, 1 October, 13 November 1913, and 14 April 1914; EEF, *Executive Minutes*, 13 October 1913; EEF Archives, I(4)2 and 12; J.R. Richmond (Weir's), *Some Aspects of Labour and its Clashes in the Engineering Industry* (Glasgow, 1917); Zeitlin, *Thesis*, pp.420–9.

107. R. Hyman, *The Workers' Union* (Oxford, 1971), pp.38–45, 48–61; J. Hinton, *The First Shop Stewards' Movement* (1973), p.219.

108. Yates, *op. cit.*, p.117; *Wages Census*, 1906, Parliamentary Papers 1911, LXXXVIII, pt.I; TUC Joint Committee, *op. cit.*, pp.11, 73–5.

109. Board of Trade, *op. cit.* (1915); Zeitlin, *Thesis*, pp.360–2, 419.

110. Wigham, *op. cit.*, pp.121–4; Jefferys, *op. cit.*, pp.218–22; for the text of the York memorandum, see Wigham, *op. cit.*, Appendix F.

111. F. Carr, *Engineering Workers and the Rise of Labour in Coventry, 1914–39*, PhD thesis, Warwick, 1979, chs 4–5; R.C. Whiting, *The View from Cowley* (Oxford, 1983); W. Lewchuck, 'Fordism and British Motor Car Employers, 1896–1932', in Gospel and Littler, *op. cit.*; S. Tolliday, 'Militancy and Organisation: Women Workers and Trade Unions in the Motor Trades in the 1930s', *Oral History* 11 (2) 1983; *idem*, 'The Failure of Mass Production Unionism in the Motor Industry, 1914–1939', in C.J. Wrigley (ed.), *A History of British Industrial Relations, 1914–1939* (forthcoming, Brighton, 1985); and my 'The Emergence of Shop Steward Organisation and Job Control in the British Car Industry', History Workshop Journal 10 (1980).

112. Zeitlin, *Thesis*, pp.466–72.

113. A more extended answer to this question will be found in Zeitlin, *Thesis*, ch.8.

Index of Names and Subjects